The FIRST-TIME HOMESTEADER

The FIRST-TIME
HOMESTEADER

A COMPLETE BEGINNER'S GUIDE
to STARTING *and* LOVING YOUR
NEW HOMESTEAD

RAISING CHICKENS • ANIMAL HUSBANDRY • GROWING A GARDEN

HOME DAIRY PRODUCTION • BEEKEEPING

RESOURCEFUL LIVING • HOMESTEAD KITCHEN SKILLS

Jessica Sowards

of *Roots & Refuge Farm,*
author of *The First-Time Gardener: Growing Vegetables*

COOL
SPRINGS
PRESS

Inspiring | Educating | Creating | Entertaining

Brimming with creative inspiration, how-to projects, and useful information to enrich your everyday life, Quarto.com is a favorite destination for those pursuing their interests and passions.

First Published in 2022 by Cool Springs Press, an imprint of The Quarto Group, 100 Cummings Center, Suite 265-D, Beverly, MA 01915, USA. T (978) 282-9590 F (978) 283-2742 Quarto.com

Cool Springs Press titles are also available at discount for retail, wholesale, promotional, and bulk purchase. For details, contact the Special Sales Manager by email at specialsales@quarto.com or by mail at The Quarto Group, Attn: Special Sales Manager, 100 Cummings Center, Suite 265-D, Beverly, MA 01915, USA.

26 25 24 23 22   1 2 3 4 5

ISBN: 978-0-7603-7235-7

Digital edition published in 2022

eISBN: 978-0-7603-7236-4

Library of Congress Cataloging-in-Publication Data available

Design: Tanya Jacobson, jcbsn.com

Cover Images: Jessica Sowards

Page Layout: *tabula rasa* graphic design

Illustration: Erin Lowe Illustration

All photos by Jessica Sowards except pages 128 (center right), 130, 131, 132 (top), 134, 140, 142–145, 149, 153, 162–165, and 178 by JLY Gardens; pages 150 and 161 (top) by Mackenzie Evans; pages 132 (bottom), 133, 135, 137–139, and 141 by Susan Brackney; and pages 53 (top), 66, and 67 from Shutterstock.com.

Printed in USA

For Sweet Miah, who is the trellis that
supports this wild and rambling dream

CONTENTS

Introduction
PG. 8

CH 1
PG. 10
Making a Plan

CH 2
PG. 30
Yard Birds

CH 3
PG. 50
Meat in the Freezer

CH 4
PG. 76
The Home Dairy

CH 5
PG. 104
The Homestead Garden

CH 6
PG. 128
Keeping Bees

CH 7
PG. 146
Resourceful Living and Natural Remedies

CH 8
PG. 166
Homestead Kitchen Skills

Conclusion

PG. 184
A Simple and
Beautiful Life

About the Author
PG. 186

Acknowledgments
PG. 187

Index
PG. 188

Introduction

The desire for a homestead was born early in me. As a child, I'd been wooed by the steady rock of a horseback ride and had conceived the notion of a farm with a red barn, yard birds, and neat garden rows. It was a poetic desire, the romantic hope of a child. I didn't grow out of it, though. I carried that dream well into womanhood and labored over it until I held it in my hands and other people could see it, too.

Our first farm was built on a 4-acre (1.5 ha) piece of ground in the middle of the woods in Central Arkansas. It was rocky land with little to no topsoil, but it was what we could afford. We learned to garden there, to build, to be homesteaders. After 8 years, we dug up the root system of our family, drove 11 hours through the southern United States, and planted ourselves on 27 acres (11 ha) of

raw land in South Carolina. We wanted to put down our roots in a new community and to build a farm from scratch. In the process of doing so, I wrote this book.

In the years that have passed since my homesteading dream budded and grew into reality, I have watched an incredible transition take place in our culture. When I first began learning about chickens, foraging, sustainability, scratch cooking, gardening, and the like, I would gush with passion to anyone who would listen. I'd watch them nod politely with glazed-over eyes, and when they changed the subject, I wouldn't bring up my desire to homestead again. Somewhere along the way, though, the tide turned.

Growing food in your backyard stopped being such an obscure idea. The conversation about food sourcing got louder, and mainstream shelves touted locally grown produce and pastured beef. Farmers' markets boomed, community gardens took off, and the back-to-basics lifestyle was validated in the mainstream by celebrities keeping chickens and starting small farms. And though I saw this happening before 2020, the onset of COVID and the subsequent lockdowns and shortages set these changes in concrete. Homesteading as a worthy way of life was no longer in question.

I receive hundreds of messages a day, and there is a repeating theme in many of them: Folks are leaving cities and buying land for the first time, and they want to know where to start. Those who aren't leaving, either because logistics won't allow it or they don't want to give up their suburban lives, are asking what they can do to grow food where they are. All over, in one form or another, people en masse are becoming homesteaders. This thrills me, and it is the reason I am writing to you right now.

There's no way I could teach you everything you need to know in 48,000 words. Entire books have been written on the topics I'll touch on in a single paragraph. This is not meant to be an exhaustive text. Rather, it is a love letter to the homestead and a guide to getting started for the homesteader, embellished with the mistakes I've made and the lessons I've learned along the way. Even if I had a million pages to fill, I couldn't teach you everything you need to know to homestead well. I'm afraid some lessons are only learned by experience. My hope, however, is to equip you for the journey, to give you a compass to navigate this beautiful way of life, and to blow a breath of inspiration on the sparks in your heart that (if you're lucky) will fuel a passion great enough to make a big leap and chase a worthy dream.

CHAPTER

1

Making a Plan

I don't want to talk to you about a trend or a fad. I want to talk to you about a worthy lifestyle. I find language fails me when I try to explain what we do. The word *farmer* evokes imagery of commercial agriculture, massive herds of animals, and crops by the trailer load. The word *homesteader* recalls ladies in prairie dresses churning butter in a one-room cabin. Even the newer term *hobby farmer* fails me by conveying I might just be one with a petting zoo in my backyard.

We are a little bit of all these people, but none of these definitions describes us exactly.

This morning, my husband, Miah, and I woke up before the sun. I brewed a pot of coffee in the French press while Miah broke the multicolored shells of our farm-fresh eggs and dropped the golden yolks and crystal-clear whites into a sizzling puddle of melted lard in a cast-iron pan. That pan was a gift to myself when I moved into my first apartment, nearly 20 years ago. I was just beginning to know the importance of sourcing food well.

I skimmed the cream off the top of a jar of milk and stirred it into my coffee, watching the dark brown give way to rich swirls of our cow Hope's great gift to us. I then cut hearty slices of yesterday's sourdough loaf and put it in the toaster. It popped up, golden and warm, just in time to be topped with perfectly fried eggs. Miah dropped a handful of herbs on top of the stack with a sprinkle of chunky salt and broke the yolks that spilled out like the sunrise, and we tucked in.

These are the humble offerings of our life. When people ask what we do, the first thing I tell them is that we eat good food and we grow it ourselves.

After breakfast, I pulled insulated coveralls over my pajamas, shoved a knit cap over my tangled hair, and grabbed the milking pail from its place by the front door. Before any other chore, we visited Hope. The family cow is the queen of the farm, and daily, I pay homage. Kneeling by her side, I breathed warm breath onto my hands and then engaged in the intimate act of hand-milking. She gave me 2½ gallons (9.5 L) of fresh milk this morning, her normal contribution. We poured it

"First comes thought; then organization of that thought, into ideas and plans; then transformation of those plans into reality. The beginning, as you will observe, is in your imagination."

—Napoleon Hill

ABOVE: *Milking is a routine that anchors life on a farm. One dairy cow produces gallons of fresh milk daily.*
OPPOSITE: *Chickens are the gateway animal on the homestead. Collecting eggs is a chore anyone in the family can do.*

into jars in the kitchen, and hours later, I smiled when I opened the fridge and saw 2 inches (5 cm) of slightly yellow cream floating atop the milk.

After milking, our chores were calling. We moved the electric fence for our pastured meat chickens. I fed my goats and slid my hand across their tightening bellies to feel this spring's goat kids twitch and roll. Miah carried a bucket of kitchen scraps to the hog pen, dumped it over the edge of the fence, and shook his head as their demanding squeals turned to smacking and slurping. I gathered eggs from the nest boxes of the laying flock. When I came across one that was still warm, having just been laid minutes before, I pressed the smooth, green

shell to my cheek until it grew cold in my icy hand. When people ask what we do, I tell them we get romanced daily by the steady rhythms of living things.

We walked through our expansive garden space and discussed the plans to prepare it for spring. There are fruit trees to plant, greenhouses to build, and soil to amend. A large plot in the pasture is cleared and leveled, and our red barn will stand there before winter is over—before kidding season begins. All around us are things to do, to build, to grow. The to-do list is long. On a good day, we tick tasks off it. On a bad day, we chase escaped pigs in the rain and our to-do list remains untouched. When people ask what we do, I tell them we work very hard and very steadily.

Can You Be a Homesteader?

The *Oxford English Dictionary* defines *homesteading* as "life as a settler on a homestead." A *homestead* is defined as "a farmhouse and outbuildings." This, I think, falls woefully short of describing this morning, all my mornings, and all my days. I like to use the term *modern homesteading*. Though I'm no scholar, if I had to write a definition, it would look like this:

modern homesteading (verb): the act of living lightly on the land, of seeking sustainability, and of growing food in a modern world; involved in this lifestyle are conscious consumption, awareness of the partnership between humanity and the earth, a reverence for the beauty of life, the embracing of a slower pace, and a desire to eat well and steward well

I think that covers it, more or less.

If you'll notice, my definition of modern homesteading does not mention the need for acres, experience, or any certain animal or skill. Homesteaders exist in all walks of life in our modern world. They live in apartments, shopping at farmers' markets to preserve produce, growing container gardens on their patios, and tending a worm farm in their kitchen for compost. They are in rental houses, growing tomatoes in raised garden beds and gathering eggs from their small, movable chicken coop. They are on a few acres in the country, expanding efforts and filling their freezers with home-raised meat. They are living off the grid in remote places, embracing slow living and rarely frequenting the grocery store.

Becoming a modern homesteader is not something that happens in a moment. In my experience, it was a decision made once, then repeated over and over in the face of failures and more convenient avenues. It is a rewarding life that costs something in terms of determination and

The homesteading lifestyle is a choice that is made day after day.

Our pastured layer flock scratches through the field around the barn, cutting down the insect population and getting extra nutrition to lay delicious eggs.

hard work but one that pays big in rewards. This journey begins long before you break ground on a farm of your own.

If I had to look at the time line of my own homesteading journey and define a starting point, the place where I really began making progress toward the desire for food sustainability, I would point to a bramble of wild blackberry bushes. I'd been reading books and dreaming of a little farm for years, wishing I lived in the country, wishing I had more money, wishing for a life that I couldn't simply produce. I'll never forget my friend Becky calling me one day and asking if I wanted to go forage wild blackberries on her father-in-law's 40 acres (16 ha). A light bulb went off in my head.

I couldn't afford much at the farmers' market, didn't have permission for chickens or even a compost pile in my rental home, and felt defeated on a regular basis by the way my circumstances seemed to render my dream hopeless. But free wild blackberries? They were a door. I picked those blackberries with a vengeance, sometimes driving to the land a few times a week. I'd bring home gallons, with my arms scratched raw and my heart fit to burst with pride. I learned to can blackberry jam, to bake blackberry cobblers, to soak the berries in salt water to pull out any hiding bugs, and to freeze berries to use during the winter months.

I realized I may not be able to buy a farm, but I had a kitchen, and surely kitchen skills were important for a

Foraging for berries on roadsides was my first breakthrough to food security.

homesteading life. I pulled my first cast-iron pan from the cabinet, where it had been sitting orange with rust. I learned to season it. I started buying whole chickens from the store and learned to break them down into parts, and after roasting one for dinner, I'd save the bones in a freezer bag. When I'd collected enough, I learned to make bone broth.

Suddenly, homesteading didn't seem so unreachable. I like to tell people, "Turn your waiting room into a classroom." When you have an unfulfilled dream, don't just sit twiddling your thumbs, waiting to get started. Prepare yourself with knowledge, right now. When you embrace the role as student in this classroom of life, lessons present themselves to you in abundance.

Circumstances do not make a homesteader; choice and determination do. I'm so encouraged to have you with me in the pages of this book as you make the choice to pursue a more sustainable and self-reliant way of life.

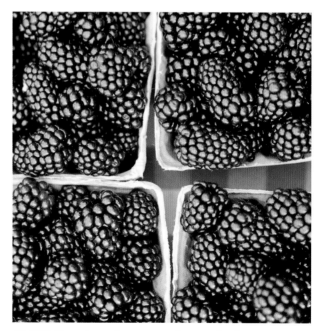

ABOVE: Shopping at farmers' markets teaches how to eat locally and value real food.

OPPOSITE: Caring for farm animals is a mutual exchange. We give them a good life, and in return they give us food security.

Creating spaces you enjoy is part of successful homesteading. Find rest and joy in the process of food growing, and you will be much less likely to burn out.

Homesteading in the 'Burbs

When the desire for food freedom springs up in your heart, living in town can suddenly feel extremely restrictive. Some suburban areas are much more forgiving of one's desire to garden and keep animals. Check with your city permit office to learn what is allowed where you live. Homeowners associations may also have restrictions that must be considered.

Many municipalities, however, allow keeping backyard hens, gardening, and even keeping a couple of small-breed dairy goats and a limited number of beehives. Don't assume that you are without options just because you are currently living in town. Your homestead may simply look different than another.

Consider any legal restrictions before purchasing property. Bylaws may restrict keeping swine or limit the number of animals you can keep on a given piece of land.

Consider the climate of any area you are thinking of home-steading. Food can be grown in most places but may require structures like high tunnel greenhouses.

What to Keep in Mind While Land Shopping

There comes a point in many homesteaders' journeys where they take the leap and move from the city to a piece of land to carve out their dream on a larger scale. Our first farm was a foreclosed home on 4 neglected acres (1.5 ha), largely chosen because it fell within our very small budget. After 8 years of growing there, we found ourselves shopping for land again to expand, eventually landing on 27 raw acres (11 ha) with no home or utilities. Here, we were heavily banking on our learned skills to build a farm from scratch.

Land shopping is a thrilling and daunting endeavor. Following are some things to keep in mind when shopping for your land:

Your goals: Before you ever look at a single real estate listing, know what you're hoping to achieve on the land you buy. Are you hoping to grow a large garden, keep bees, raise dairy animals, raise meat? Are you planning on having a market garden or offsetting your costs by selling cut flowers? Do you homeschool your children or commute to work? Are you hoping to remodel a home or build one? Will you be hiring out work or doing much of it yourself? All of these factors will weigh into your

chosen land, and it's good to have your goals in front of you before you even start looking. Make a list of goals and be realistic with your time line.

The laws and restrictions on the land: Just because a property is rural, large, or even outside of city limits does not mean you'll have free rein to do whatever you please with it. Before you set your heart on any piece of land, dig into the laws and restrictions that govern it. I knew a homesteading couple that bought 11 acres (4.5 ha) that had a clause in the contract denoting a restriction against raising swine. The seller lived next door and didn't want to smell pigs, so he'd worked it into the agreement. The homesteaders were so excited to have found a piece of land they liked, they dismissed the limitation; but years down the road, they discovered that the rocky, wooded lot wasn't great for some of the other protein sources they had originally planned to raise. Pigs would have been well suited to their homestead, but they legally weren't an option.

Before you sign on the dotted line for any property you will be sinking resources into, understand the options you have with that property. Find out if there are any restrictions by a homeowners association, clauses from the seller, or government laws that might hinder you from fulfilling your homestead dream.

The laws that affect your lifestyle: If you are searching for a property in an unfamiliar region, make sure you understand the laws existing in that region that will affect your family and lifestyle. Understanding home-schooling laws, tax rates, and the freedoms afforded to you in a certain place should weigh heavily on a choice to relocate. If you own a business or work in a certain field, job availability should be explored. Talk to local people in the region you are considering and try to get a feel for the culture.

Readiness of the land: Understand the cost of preparing wild land. Raw land is usually much cheaper than an already-cultivated property, and there is a reason for that. It is expensive and time-consuming to turn raw land into a working homestead. I've talked to many people who bought raw land outside of the city limits and cultivated it over the course of years while they contin-ued to live in town. Before you make any deals for a piece of raw land, get some quotes for the jobs that will be necessary for you to really use your land. If you need a driveway or trees cleared or a well, make sure you understand what they will cost. Also consider how much labor your family can commit to the building of a homestead. Will you or your partner be working away from the home? Raw land can be a beautiful opportunity to build something custom, but you need to understand the commitment before you take the leap.

The neighbors: You can't always choose your neighbors, but when land shopping, there may be some deal breakers in proximity. If your goal is to grow an organic farm, consider the nearness of any commercial agricul-ture. If your prospective property is adjacent to any large conventional ag fields, runoff and drift of chemical pesticides and fertilizers may dash your dreams.

Water and drainage: Having a source of water, with water rights, is a great perk to a homestead property. Pumps can be used to irrigate gardens, and depending on your situation, animals may be given access to ponds and streams. If you can observe the land during a wet season, this is ideal. Understanding where water pools and how it drains is imperative to farm planning. At the very least, check public records to know if your prospective land is

Successful homesteading is a family affair. Homeschoolers should check local homeschooling laws before relocating.

in a flood plain. If it is, flood insurance is expensive, and you will be limited on placement of structures. If a property is large and a small part is in a flood plain, it's not a deal breaker; but if you won't be able to use much of your land for barns, a garden, and the like, it would probably be best to keep looking.

Soil: Good soil can be built. Our first farm was on a rocky, wooded ridge, and the topsoil was virtually non-existent. We successfully gardened by bringing in compost, layering organic matter, and spending years building soil. It was a lot of work, and a nice layer of rich soil was high on my priority list when shopping for a forever farm. Depending on the region and history of the land you have access to, finding a property with ready-made soil may be impossible. I suggest taking a shovel to any land you are considering buying. Dig samples in multiple areas. Get a soil test. Since this can be amended, it may not make or break your real estate deal, but it will be good to know what you're working with from the start.

Knowing Your Why

Homesteading is hard. Animal husbandry, gardening, and the required discipline to push yourself beyond the confines of your usual comfort are learned skills for most of us in the modern world. Sure, it's a beautifully romantic life. Sure, a lot of people are realizing the value and necessity of closing the distance between themselves and their food. Obviously, I believe in the value of homesteading or else I wouldn't be writing this book.

However, there are going to be times you want to quit. I heard an old farmer say once, "Where there is livestock, there is dead stock." He was right. That quote is a solid, unarguable truth, and still it provided no comfort to me as I sat on the barn floor with the body of the first goat I ever lost. Her name was Delilah. We'd done everything we could to save her, but she still died.

This one-month-old Jersey calf, Halle, was the first calf born on our farm.

Homesteading means long days and physically demanding work, and even if you do everything right, sometimes you still fail. Since we started our first farm, I have cried, bled, and puked doing farm-related tasks more times than I can count. I've been bruised and injured and had to push through chores while extremely sick. I've laid in bed at 11 o'clock at night and felt the blanket of exhaustion fall on me like a thousand pounds, only to remember one last task and drag myself back up to administer medicine or feed a bottle or any number of things that cannot be postponed until morning.

I have wanted to quit, over and over. Even now, I know there are moments in my future where I will face heartache and must fight down the urge to give up. I won't give up, though, because I know why I'm doing this.

I keep going when it's hard because an inexpressible sense of pride fills me when I see jars of preserved harvest lining the pantry shelves or when I stand over the freezer after a long day of butchering this season's meat chickens. I push beyond failures because whenever the news reports food shortages or impending winter storms cause the grocery store shelves to be wiped clean, I do not feel afraid. The hard work is worth it when I watch my cows graze green pastures or my chickens chase bugs in the yard, and I know that no living creatures are suffering miserable lives to feed my family. And though there are certainly moments of discomfort, I do not quit because I also notice when I pick up a feed bag or squat in the garden to pull weeds that this life has made me stronger. All these things pale in comparison to my biggest reason to persevere, which is my dream of building a legacy and making a route for others to live a life uncommon.

These are my reasons why, and with them comes the mighty courage brought on by deep convictions. I often hear homesteaders talk about the difficulty of the lifestyle to fledgling wannabe farmers with a tone of encouraging them to quit now before they get in too deep and lose too much. It reminds me of an adult eating a chocolate bar and telling an onlooking child, "You don't want any of this. It's yucky."

Homesteading is wonderful, worthy, and truthfully, sometimes yucky. It's hard, but you can do hard things.

ABOVE: Raising my children to understand stewardship of land and animals is a big factor in my "why."
OPPOSITE: Growing your own food creates a great sense of pride.

I won't sugarcoat this, but I also won't discourage you from it. Know your why. Write it down now, while you are still ignorant to the pain and the cost of pursuing your dream life. Muster the courage of your convictions and move forward knowing that some of the most beautiful results come from heartbreaking, backbreaking, and altogether worthwhile work.

Goats can be incredibly destructive to the vegetation. This makes them perfect for clearing brush, but they should be kept away from orchards and gardens.

Laying Out the Farm

This is the part of the book where I tell you exactly how to lay out your property for maximum production and enjoyment and least labor and worry. I'm just kidding. I would like to do that for you, but there is no one-size-fits-all approach to designing your property. You may live in a suburban neighborhood, or you could have 50 acres (20 ha). These places will look very different. Even two farms of the same size will look different based on your needs and wants and the features of the land.

As you're starting the process of laying out the farm, remember your why. Your property will look different if you're growing food for your vegetarian family of two versus if you are producing food for an omnivorous clan of six. As homesteaders, you have the flexibility to do what works best for you.

Get to know your property for a year or so before making permanent changes. You'll learn something about your property in every season that will help you make better decisions. You even may be able to get details about the land from a real estate agent, the former owners, or a trusted neighbor.

While scoping out the property in this first year, watch how water moves, where water drains or pools, the best sunlight and shade, wildlife patterns, and the like. Look at the vegetation already growing there. Research those plants, their growth habits, and their soil needs, and you'll learn a lot about your land. Do soil tests. Consider the assets already on the property, such as an existing home or barn, a driveway, a flat site appropriate for building, electrical service, or a well. On our South

Carolina farm, we were blessed with two ponds, some timberland, and areas with several inches of rich soil due to years of cattle farming. These helped dictate our uses of the land.

For a guide to any homestead layout, you can look to the permaculture principle of zones. Permaculture is a system of living in tandem with nature, sometimes described as permanent (perma) agriculture (culture). In this system, a set of physical zones is defined, from 0 to 5, with 0 being closest to human activity and 5 being farthest. The idea is to organize your farm so that the elements that need the most attention and require the most energy are closest to your center and those that need the least attention and require the least energy are farthest away. It makes sense to put something you deal with every day just outside your back door while putting something you only need to think about a few times a season at your farthest reach.

Your farm layout might start with a list: What do you want to do here? Maybe this includes a vegetable garden, fruit trees, laying hens, dairy goats, pigs, a greenhouse, and a sitting and play area for your family.

Then get a general idea of your zones by drawing concentric circles, from your home outward. On a small property, you might not have zones 4 and 5, and that's okay. It's also likely your home is not in the physical center of your property, and that's okay, too. On our farm, the garden belt—the vegetable-growing space that runs along our driveway—is in the middle of the property. It's close to our home and far from the wildlife that lives in the timberland and tree lines. Looking at the garden belt spatially from the house, the annual plants are closest to the house, and the fruit trees are farthest from the house. Pasture lies beyond that.

Now plot your farm activities by zone so you can move through and work on your property using the least amount of energy and time. Look at what you'll spend the most time and energy on each day, look at the physical attributes of your property, and see how close you can put that to zone 1. Think about it this way: Spending 10 minutes walking from one place to another twice a day to complete a task may not seem like a big deal right now, but at the end of the week, that's more than 2 hours that you spent walking for that one task. At the end of the month, that's 10 hours that you could have spent doing anything else.

In this layout, also consider how the assets from each zone can work for the zones next to them. Water runoff

Areas that are frequently visited are considered zones one or two. A greenhouse for seedlings needs to be tended daily, and therefore, should be located near the home.

Permaculture Zones

Zone 0
Your home or other gathering point and center of activity, such as a barn or shed. Humans are the primary consideration in this zone.

Zone 1
Immediately outside your home. You might think about a small kitchen garden or herb garden here. This is probably where a sitting area for your family, a play area for your kids, and a clothesline make sense.

Zone 2
Still close to your center, but maybe you put on shoes to get this far. This could be your larger vegetable garden, greenhouse, or high tunnel.

Zone 3
It's not inconvenient to get to zone 3, but neither is it right outside your door. Perennial plants, fruit and nut trees, the equipment shed, and livestock barns might be located here. Zone 3 is considered the edge between the cultivated area and the self-managing land.

Zone 4
Your presence is less here. You do more passive observation than active work in this area, as it's largely self-maintaining. This could be pastures and food forests. If you live in a suburban area or on a smaller property, your zone 4 and beyond may be more constrained. Homeowners associations, for example, rarely go for the wild woodland look.

Zone 5
The area farthest from your home that rarely sees human intervention. This may be a tree line at the edge of your property or another wild, unmanaged area. It may be a place where you take walks and forage but not a place where you mow or remove brush.

from your house's roof could provide water for a garden or livestock. Manure from a barnyard could be composted for the garden. Scraps from the garden could be fed to the pigs. Each zone can benefit its neighbor without too much more effort on your part.

RIGHT: A kitchen garden located near the house can be easily accessed for a quick meal.

OPPOPSITE: Keeping vulnerable animals like chickens close to the traffic of the homestead provides ease of care, and it also provides protection to the animals. Predators are more likely to attack far away from lights, noise, and movement.

Making a 5-Year Plan

From the start, I suggest a 5-year plan. This keeps you from being in a rush and lessens the overwhelm that comes with starting something new. Know that you don't have to build your homestead dream in a day or in a year. Know that you *can't* build it in a day or in a year. Allow yourself the time to develop a healthy infrastructure and a productive farm while also having a rich, full, and fruitful life. Build your farm to thrive, both in terms of food production and personal enjoyment. You'll have a place you're far more invested in.

Looking ahead 5 years allows you to think long-term but not too long-term. If there's one thing for certain in homesteading and in life, it's that things will change. Planning for 5 years gives you a direction and allows flexibility.

A 5-year view also allows you to give your property time to develop to its full potential. If you are planting perennials, you may be waiting 3 years or more to have a first harvest. If you are creating new pasture, it will take a season to become established for sustainable grazing. A 5-year plan builds in the patience to see these through.

Start your plan by going back to your reason for homesteading. Set goals that incrementally point to your purpose for being here. Make these goals small and attainable so you can gain confidence and motivation to keep going. Consider what each goal will provide for your food production and your quality of life and set priorities from there.

With goals in place, break down each with tasks to complete and skills to learn. Homesteading offers endless opportunities to gain new skills. Be realistic in allowing yourself time to learn and to do. These things often take longer than you expect, and it is easy to forget just how much goes into setting up a homestead. It is hard for me to tell you to take it slow because I have big dreams and big plans. The truth is, homesteading is a journey that will take a good, long while. Focus on the reason that you're living this life in the first place and let your 5-year plan guide your next step.

If you are still in the waiting room of your homesteading dream, build that time into your 5-year plan. There is value in gathering a skill set before you need to put it to the test, just as I did with my kitchen skills years ago. Years 1 through 4 of your 5-year plan could be spent on education and experiences, while year 5 might be buying property, making observations, and doing future planning. There is no one right way to plan for your homestead.

Put this 5-year plan somewhere you can see it. Post it on the refrigerator, above your potting bench, or in your feed room. Let it be a reminder of how far you have come, the reason you're living this lifestyle, and the small step you'll take next.

Planning to live your homestead dream is a beautiful and overwhelming process. Sometimes you have to be brave enough to dream and trust that this will work out. The first all-homegrown crisp salad that you serve to your friends around the dinner table, the first just-laid egg that you fold into a loved one's from-scratch birthday cake, the first winter day that you open your pantry to pull out a jar of red tomatoes that you grew and canned yourself, these moments make it all worthwhile.

OPPOSITE: This is Honor, a Saanen doe. We brought our first goats home two years into our homesteading journey. They were part of our first five-year plan.

CHAPTER

2

Yard Birds

Yard birds are the gateway animal to the homestead life. Countless city dwellers were lured into the idea of a few cute chicks thinking maybe they'd just be productive pets, but then the eggs came. Glorious, fresh eggs showing up in your backyard every day, when you are accustomed to things like this only coming from the grocery store, is a revelation. The shells are stronger, and the yolks are richer. The eggs actually taste like eggs. Suddenly, you find yourself considering other ways to create food sustainability.

I will warn you of a phenomenon called "chicken math," often joked about in the world of backyard chicken keepers. You start with a tiny flock of chickens, and within a year, you've gained so many birds that you're lining up egg customers and tucking eggs under a broody hen to grow the flock again. Chickens are unbelievably addictive creatures.

They are work, yes, but easy to keep and enjoyable to watch; and they produce a steady stream of food. Chickens also reproduce quickly and provide nitrogen-rich manure that, with a bit of time in the compost pile, can cycle back into nutrition for your garden soil. They can be kept in small urban and suburban yards, and even children can care for them. Chickens are a great place to start for the aspiring homesteader.

"There are days when I am envious of my hens: when I hunger for a purpose as perfect and sure as a single daily egg."

—Barbara Kingsolver

ABOVE: Daily fresh eggs give many new homesteaders their first taste of food security.

OPPOSITE: A fertilized egg requires twenty-one days of incubation to hatch a new chick.

Building Your Flock

Despite the appeal of chicken math, more chickens are not necessarily better. Just how many chickens do you need? The most productive laying hen will produce an average of five eggs per week for most of the year for the first 3 years of her life. Decide on the number of eggs that you want for your homesteading goals and look for the chickens you need to fill that number. Keep in mind that egg production slows down in the hottest part of summer and as the days get short going into fall and winter. The eggs will come in considerably faster as the days get longer in the spring.

Plan for a flock no larger than you can manage within your production ideals. We feed non-GMO (non–genetically modified organisms) feeds and treat the chickens with herbs and nonchemical medicines, if possible. We like to handle our birds when they're young so they're docile and friendly. These things are easier to do with a smaller flock.

Also consider the number of chickens your coop and yard can handle. Adequate space is essential for bird health and egg production, not to mention their quality of life.

TOP: A Bielefelder hen provides five large brown eggs a week.

ABOVE: Our docile Bielefelder rooster, Elvis, provides protection for our ladies and gives us the option of hatching fertile eggs.

When starting your flock, the question of which came first, the chicken or the egg, is less relevant: You start with a chicken. The question of adult chicken (hen) or baby chicken (chick) remains. There are reasons a first-time homesteader could want either, and it comes down to the time and effort you can invest.

Hens cost more to purchase, though you'll get eggs sooner, so you're putting less money into the feed it takes to produce each egg. On the plus side, you know you have a hen and not a rooster, which is harder to tell when buying chicks. On the downside, you risk getting a bird that's actually older than advertised, and you risk getting one that's sick and could infect the rest of the flock. That last consideration is a big one that we learned early on when we unknowingly introduced a disease to our homestead.

Chicks are inexpensive, but they're fragile. For the first few weeks, chicks can't regulate their own body temperature. It's up to you to keep them safe and warm in a brooder area until they have enough feathers to live outside, which is generally around 5 weeks of age. You get to experience the joy of watching fluffy, tiny chicks grow into silky, full-grown hens. They'll be 18 to 26 weeks old before they start laying.

Chicks are readily available at feed stores come spring-time, and hatcheries can even ship day-old chicks to you in the mail. The folks at the post office will never forget you once you've picked up a box of peeping chicks.

Starting with chicks from a reputable source, you know they're healthy. Most mainstream hatcheries are part of the National Poultry Improvement Program, which certifies their birds are free of certain diseases.

While you can buy chicks from a hatchery already sexed, there is a small chance you'll end up with some roosters. Some breeds are sex linked, meaning the female chicks (pullets) look different than the male chicks (cockerels). Besides those markers, it's understandably difficult to determine a chick's sex.

Hatcheries usually have a minimum number of chicks they will ship. If that minimum happens to be more than the number you're permitted to keep by your municipal ordinances or more than your homestead can handle, split your order with a friend.

A barnyard mixed-breed hen lays well and has the rigor of breed hybridizing.

If you choose to order through a hatchery, use the catalog as a reference tool. It has everything you need to know about each breed, like skin color, comb type, egg production, mature weight, climate preference, and more. The downside to looking through these catalogs is that it might result in a large chicken order. There's that chicken math again.

Murray McMurray Hatchery in Iowa is my favorite mail-order source for chicks. It's a small, family-owned hatchery and the owners pretty much have a database of chicken facts in their heads. They work with The Livestock Conservancy on maintaining rare and heritage poultry breeds, which make wonderful homesteading birds that aren't used in commercial flocks.

Children thrive with responsibility on a homestead and caring for a chicken flock is a great place for them to help.

Chicken Breeds

Making the decision to bring chickens to your homestead may be easier than making the decision of which chicken breeds you want. Back to that chicken math, it's not just the numbers that tend to multiply; it's the breeds, too. There are so many interesting chickens.

Like lineages of seeds, lineages of chickens have been bred as hybrids and heritage. Hybrids are not "genetically modified;" they're simply different breeds crossed to create a chicken with a certain trait, like maximum egg production or cold hardiness. There are reasons to want hybrids and reasons to want heritage breeds.

To start, think about egg production and mothering traits. I like to keep some broody hens who will sit on and hatch out a clutch of eggs and some steady layers who are not interested in the work of incubating eggs. Broodiness has largely been bred out of hybrid chickens, but hybrids often lay eggs sooner and more frequently. In heritage breeds, mothering traits like broodiness ensure the sustainability of your flock. Assuming you have a rooster, broody hens mean new chicks. New chicks mean always having new layers to continue the flock when loss occurs due to age or predators. The downside to having a broody hen is that once she's laid her clutch of eggs, she'll stop laying because she turns her attention to incubating her eggs instead.

Feed consumption is another consideration. Heritage chicken breeds tend to be better foragers than hybrid chickens. Kept outdoors, they'll readily snack on insects, grubs, seeds, and weeds to supplement their diet and add nutrition to their eggs. Heritage birds will also roam farther from home in search of snacks. Hybrid chickens may have better feed conversion, meaning that while they aren't as interested in snacking in their yard, they require fewer calories from their feed to produce an egg. They're also more food motivated and tend to be less brave about leaving their area for foraging.

Your climate should play into your chicken choices, too. Any chicken breed can live in any climate with the right care, but not any breed will thrive in any climate. Smaller-framed chickens with larger combs are better suited to very hot places. These are often Mediterranean breeds, like Minorcas and Andalusians. In a very cold place, consider a breed with a smaller comb, which is less susceptible to frostbite. Those with feathered feet will also do well. Wyandottes and Brahmas are two good cold-climate chicken breeds. If you're keeping hens only, you have more options, as hens have smaller combs than roosters. The single-comb hens—like Black Australorps, Plymouth Rocks, and Rhode Island Reds—do fine in cold climates, but the roosters suffer.

We keep separate flocks of layers and meat chickens, but if you're looking for hens that both lay eggs well and have the body type of meat birds, dual-purpose heritage breeds are the way to go. Orpingtons, Delawares, Cochins, Plymouth Rocks, and Rhode Island Reds are dual-purpose chickens made for the homestead.

Overall, the chickens that I like best for homesteading are the heavy-bodied types, which are hearty and versatile. These breeds can tough it out through inevitable beginner-chicken-keeper mistakes in most climates in the United States. I also want chickens that add beauty to the farm and color to my egg basket. It may seem like a lot to ask, but there are so many breeds out there that meet our needs, I still have trouble choosing.

A flock of twenty-five chickens can produce a staggering number of eggs daily.

With that said, here are some of my favorite homesteading chicken breeds:

Barred Rocks: A variety of Plymouth Rock chickens, these are productive brown-egg layers that have been around since the mid-1800s. They're hardy, dual-purpose heritage chickens with a single comb. The "barred" in the name refers to the beautiful white stripes on their black feathers. Other varieties of Plymouth Rocks have different coloring. Expect as many as 280 eggs per year from one Barred Rock hen.

Black Copper Marans: These chickens have a slimmer, more upright body shape than your typical farm chicken, and they're not often found on small farms in the United States. I love them for their dark-chocolate-colored eggs. They'll typically lay 200 or more eggs per year. For a fun home-breeding experiment, cross Black

Copper Marans with a blue egg layer for a chicken that produces green eggs.

Bielef elders: Bielef elders are beautiful chickens and a relatively new breed. I can't explain how friendly these gentle giants are. The hens can lay up to 280 brown eggs per year and begin laying at an early age. The roosters weigh 10 pounds (4.5 kg) when mature. As chicks, they're easily sexed, as cockerels have a yellow dot on their heads, while pullets do not.

Ameraucanas: Miah says Ameraucana chickens are bougie because many of them have beards, they have fluffy butts, and they come in all different colors. We love them for the accent they add to the flock and for their beautiful, blue eggs. They are medium-size, hardy with pea combs, and a bit flighty rather than broody. Their range of feather colors includes blue, black, blue-wheaten, brown-red, white, wheaten, buff, and silver. Ameraucana hens will lay around 200 eggs per year.

Orpingtons: These large, dual-purpose heritage chickens are gorgeous. They're friendly, too, and come in a range of colors. Miah's all-time favorite chicken was a Buff Orpington. Buff is the standard color, but don't forget about chocolate, lavender, and white. They'll lay eggs through the winter and will get broody. You can collect up to 200 eggs per year from an Orpington hen.

Cochins: Hear one Elvis-like crow from a massive, gentle Cochin rooster and you'll know why these are on my list. We have never had a mean Cochin rooster. Cochin hens are very broody, even to hatch turkey and duck eggs, and will lay 180 or so eggs per year. They're naturally slow moving and so are easy to handle when needed. This breed comes in a range of colors, from buff to black and golden laced to barred and colors in between.

Roosters in the Flock

Hens are obviously necessary for your laying flock. Roosters are worth considering, though you do not need a rooster to have eggs. That's a common misconception. Hens will naturally lay eggs with or without a rooster. In many urban and suburban neighborhoods, roosters (and their crowing) are not welcome. While this is a bummer, it doesn't mean you can't still have great eggs.

You do need a rooster to have fertile eggs for hatching your own sustainable flock. If you're doing your own breeding, count on one rooster for every six or eight hens. More roosters than this, and they could get territorial.

Roosters have a bad rap for being mean. We have had more nice roosters than mean ones. Some of that is the personality of the individual bird, the disposition of his breed, and how he's handled when he's young. If you start with chicks and handle them regularly, your roosters will grow up more docile.

Roosters provide some protection for your flock with their loud crowing, sharp spurs on their legs, and willingness to engage in trouble, but in the end, they can't provide all the protection needed.

OPPOSITE: A negative encounter with a mean rooster can scar a person for life. We keep roosters for sustainability, and most are very docile and enjoyable. If ever a rooster begins attacking us or our children, we quickly make a pot of mean-rooster stew.

A Home for the Flock

When we brought home our first 30 chicks in a cardboard box, we didn't have a fence, a coop, brooder lights, or anything necessary to do this right. My first piece of advice for you is to be prepared before you make your chicken purchase.

Chickens are hardy animals. They don't need a fancy coop—just a predator-proof space with the basics: shelter from the elements, roosts, and nest boxes. You can build your own coop or purchase one already made. We built one using an old dog kennel and wood pallets for less than $300, and you could do it for less if you can repurpose more materials.

You might choose to use a mobile coop. These are often called egg mobiles, chicken tractors, or "chickshaws." We built ours using plans developed by our friend Justin Rhodes. Mobile coops allow you to move the chickens around your farm, spreading out the nutrients in their waste, giving them fresh ground to forage, and allowing them to clean up parasites left behind, if they're following your livestock. These coops need to be moved every day or every few days, depending on the number of birds and time of year.

A static coop may be more suitable to your yard, space available, or time commitment. This requires bedding that has to be cleaned and kept fresh.

Assuming your chickens will have access to the outdoors most of the day, they need just 2 square feet (0.2 m²) per bird inside the coop. Give each bird 8 inches (20.5 cm) of

Chickens with larger combs are better suited for warmer climates, as cold temperatures cause frostbite in combs and extremities first. Consider pea-combed birds for cold regions.

THE FIRST-TIME HOMESTEADER

Static coops have a lovely aesthetic but can mean more work for the chicken keeper.

roosting space, a place they can sleep off the ground. Allow one nest box for every five hens.

In most cases, it's best not to put supplemental heat in your coop. It can cause problems with ventilation and can be a fire hazard. Rest assured that your chickens are fine in the cold, just like cardinals, hawks, chickadees, and other wild birds are. During very hot periods when the nights don't cool down, we might put fans in the coop to help with air circulation.

Chickens are helpless against predators, particularly at night. Predator-proofing a coop means allowing for no gaps larger than 1 inch (2.5 cm). A motivated predator could squeeze itself through a 1-inch (2.5 cm) gap or reach a paw through there to grab a bird. Lock your chickens inside from dusk to dawn to keep out aerial and ground predators.

In the field, an electrified poultry net fence is effective protection against ground predators. This fencing is easy to move and durable for many seasons. Aerial predators, like hawks, are harder to deter and require some creativity, like hanging reflective material above the chicken

Movable electric fences (powered by solar panels) and mobile housing allow for moving a chicken flock to fresh grass regularly.

run—shiny ribbon or those CDs you have no other use for—or putting out a fake owl totem. While we have had mixed results with our livestock guardian dogs, they can be useful in keeping chickens safe as well.

Feed Them Well

I've heard people say brown eggs are healthier or that they prefer the flavor of blue eggs. In truth, egg nutrition and flavor are determined by the diet of the chicken and have nothing to do with the egg color. A protein-rich diet, with a well-rounded feed and access to pasture, will result in nutrient-dense, delicious eggs with the loveliest dark-orange yolks.

I enjoy bringing garden scraps to the chickens. It's entertaining to watch them chase after a rotten squash as the squash hits the ground and breaks into a dozen flying pieces. As nice as it would be to let the chickens live off the land and our garden and table scraps, this will not provide all of the nutrition they need to be productive layers.

Commercially formulated feeds offer complete nutrition. We feed a locally milled, non-medicated layer pellet. Your local feed mill may have a grain-and-mineral mix appropriate for laying hens if you want a complete feed in less processed form.

Scratch is a mix of seeds and grains that I like to provide to the chickens as a bonus treat. It's good for entertainment and encourages them to forage by hunting and pecking the ground. Crushed oyster shell, too, provides an added calcium source for eggshell development.

Chickens need fresh water daily.

Any scraps we toss their way and the seeds, insects, and grubs they forage supplement these feeds. Chickens can eat most everything that comes off your kitchen table, except onions, citrus, and avocado skins. Chickens are omnivores—just wait until you see one catch a mouse—as much as we like to think of them as being vegetarian.

Keep their food secured because wild birds, mice, rats, and other critters love chicken feed. If the feed gets wet, it will mold and spoil. You may place a metal trash can with a tight-fitting lid in a garage, barn, or shed nearby.

In talking about nutrition, I cannot overemphasize the importance of fresh, clean water. Chickens don't like warm water and may refuse to drink it, even in hot weather. Get a larger waterer than you think you'll need and check it often.

When Production Ends

When you end up with more roosters than needed and when your hens have reached the end of their egg-laying lives, you have tough decisions to make. On a sustainable homestead, feeding animals that are not feeding you is something to question.

Hens are most productive in the first 2 years of their lives, and they'll produce fewer eggs over the next 2 years. They'll generally stop laying at 3 or 4 years old, and they could live 8 years or longer.

Your extra roosters and spent hens could live out their lives as pets on your farm, or you could harvest them to feed your family. Older chickens do not have the body type or meat quality that you're used to. You won't get a grand roast chicken meal from them; rather, they're best used as stewing hens and for stock making.

Poop Is Money

A beautiful basket of fresh eggs is the primary motivation behind chicken keeping, and the nutrient-rich waste chickens offer is a close second. Chicken poop in its fresh form is too "hot" for the garden and can damage plants and roots. It also can pose a food-safety hazard to fresh food. Composted, it's a great source of organic matter, nitrogen, phosphorus, and potassium.

One benefit of keeping your chickens in a static coop is the ease of collecting manure for composting. The coop needs bedding of some kind. We've used straw in a deep-bedding system, adding more bedding each week and allowing it to compost in place. We also tried sand. You might use wood shavings or another readily available organic material.

Stay on top of coop sanitation. A buildup of dirty bedding can create smell and respiratory issues. Using sand as our bedding, we had virtually no smell. It was more work up front to load in 8 to 10 inches (20.5 to 25.5 cm) of sand, but it only needed to be changed once a year. Other than that, we just sifted and broke it up as it got dirty.

The bedding mixed with the manure provides a good carbon-nitrogen ratio for your compost. You'll need to water and turn the pile to keep it hot and allow it to fully break down.

With mobile coops, you could run your chickens over the area you will use as your garden, then let it age in place. Food-safety guidelines say to incorporate fresh manure into the soil a minimum of 120 days before harvest when the food comes in contact with the soil, such as root vegetables or strawberries. For foods not in contact with the soil, like tomatoes or pole beans, it's a minimum of 90 days.

If this all sounds like a lot of information just to get started with chickens on your homestead, it is. Chickens are the easiest of the animals to keep, but they still require attention and care. Don't lose sight of the point here. There is a trade-off to waking at dawn every day to let them out, coming home at dark every day to put them inside, moving their mobile coops and fences, and ensuring they're fed and watered. In return, you are gifted the most glorious fresh eggs you've ever experienced, a great by-product for your garden, and the pleasure of having feathered friends in the yard.

A six-week-old Cornish X chicken

TOP: *Colorful eggs have the same flavor and nutrition as their brown and white counterparts.*

ABOVE: *Selling colorful eggs as hatching eggs can bring a larger profit than selling them for eating.*

Technology Tip

It's so important to keep chickens in the coop at night to protect them from predators. Being home at exactly dusk to close the door and being outside at dawn to open the door is a big task. While we still need to be out there for feeding and collecting eggs, we've found automatic coop doors take one chore off our plate. We modified both our chickshaw mobile coop and the static coop with an automatic door that opens with the daylight and closes with the dusk, letting the birds in and out at just the right times. There's no need to chase your chickens into the coop at night. They will naturally go home to roost just before dusk.

A Colorful Egg Basket

Eggshell color is determined by a chicken's genetic makeup, just like their feather color and skin color. Egg color genetics is a fascinating science, and breeders have experimented with selective breeding and cross-breeding to bring out the darkest chocolate and olive eggs as well as the brightest blues and even some rare colors, like lilac and pale pink.

These colorful shells don't add nutritional value to your eggs, but they do add beauty and interest. On a more pragmatic note, it may also make your eggs more marketable. If you are hoping to sell extra eggs at a farmers' market, a consumer is often more willing to pay a premium for the colorful eggs that they won't find at the supermarket.

In chicken genetics, the blue eggshell gene is dominant. Purebred blue egg layers, like those listed in the following table, will carry two copies of the blue eggshell gene. When crossed with a breed that lays a different color, all offspring will still have one dominant blue eggshell gene. Therefore, all first-generation offspring of these birds will lay a colorful egg. It may be a lighter blue or a shade of green, depending on the genetics they were mixed with.

Often, when searching for colorful egg breeds, you will find birds listed as "Easter Eggers." These are mixed-breed birds that carry at least one blue eggshell gene. You may also see listings for sage eggers, olive eggers,

Common Breeds for a Colorful Egg Basket

Chicken Breed	Egg Color	Egg Size
Ameraucana	Blue	Large
Araucana	Blue	Medium
Crested Cream Legbar	Blue	Medium-large
Marans (French Black Copper, Blue Copper, and White)	Chocolate brown	Large
Marans (also called Cuckoo)	Light chocolate brown	Large
Penedesenca	Chocolate brown	Medium-large
Welsummer	Brown with chocolate speckles	Large
Whiting True Blue	Blue	Large

chocolate eggers, and the like. All of these are names given to birds that came from selective cross-breeding for the goal of achieving a certain egg color. They are hybrid birds and can often be expensive, but they are a great way to add rare egg colors to your basket without having to do generations of breeding yourself.

Why Aren't My Birds Laying?

There are a number of reasons why chickens stop or slow their laying. Some sleuthing can reveal the reason, and often it's an issue related to health or management that's easy to remedy.

Mites: One of the most common reasons chickens aren't laying is because of mites, which are tiny insects that can infect your flock from wild birds. It's easy to see if a chicken has mites: Turn her upside down so she's calm. Spread the feathers away from the skin and you'll see the red, dark brown, or white bugs crawling. Keeping your coop clean and providing a dust bathing area filled with diatomaceous earth will help prevent mites.

Wet feet: Any animals kept in a small space over time will wear down the grass, and you'll end up with mud. Muddy ground leads to wet feet, which makes the chickens uncomfortable. Wet feet also leads to muddy eggs because the hens walk on the eggs in the nest boxes when they go in to lay. Keep the ground dry by moving your chickens regularly or, if moving isn't an option, lay down straw as needed to reduce mud.

Low light: While this isn't the case for every chicken, some chickens stop laying through the winter. Less than 14 hours of light per day hinders a regular laying schedule. You can put artificial light in their coop, like a string of holiday lights or a single light bulb, on a timer. This encourages them to continue laying, though it also shortens their productive life. Some people feel strongly about giving chickens a break in the winter. Hens are born with the number of eggs they will lay in their lives. By forcing them to lay through the winter, you're not going to get more eggs from them overall—just more from them in a shorter time span.

Poor feed: Sufficient nutrition is essential to keep your chickens laying. You can feed them extra protein in the form of chicken treats, black soldier fly larvae, and chicken scratch. Chickens may consume more feed in the winter to keep warm and less feed in the summer when they are foraging more. Be sure you are feeding the right amount and quality year-round.

Elvis watches over his flock, crowing to alert them to overhead predators.

Stress: Stress affects your chickens just like it affects you. Any major stress, including a new coop, new birds in the flock, or a big storm, can make your chickens lay less for a few days or even a couple of months.

Molting: Chickens often lose a lot of their feathers at the end of fall and grow in new ones and stop laying while they're at it. This process is called molting. Help them through the molt by supplementing their feed with additional protein.

Age: Hens are born with the number of eggs they will lay in their lives. If you've kept your chickens under artificial light so they lay more consistently through winter, they'll deplete their egg supply earlier in their life. Hens generally lay the most eggs for the first 2 years and slow down in years 3 and 4.

If your chickens are free range, you can never assume they have completely stopped laying. Always assume they've stopped laying where you can find the eggs. Chickens like a good egg hunt. I've found eggs in the most random of places. Especially if they're preparing to go broody and sit on a clutch, they'll go off on their own for laying. Before you panic, thinking your flock's laying days are over, do some exploring in the areas that your birds frequent to look for your missing eggs.

Fowl Friends

Guineas

The barnyard alarm system, guineas add interest in sounds and looks to the poultry yard. Helmeted guinea fowl are the most common type raised in the United States.

Guineas are known for letting out a call when anything is amiss, alerting their poultry friends to take cover when a raptor is above and startling intruders of all kinds. This enthusiastic alert presents a challenge for neighbors in proximity.

They are voracious insect eaters and can control the population of ticks, slugs, and Japanese beetles. The downside to this is that they will eat honeybees, too.

It takes about a year before guinea hens start laying eggs. They lay only in spring through fall, an average of one egg each day. A guinea egg is about half the size of a chicken egg, with a higher yolk-to-white ratio.

Guineas reach the size of meat birds, 1¾ to 2½ pounds (794 to 1134 g), at 14 to 16 weeks of age. Their meat is tender and somewhat gamey.

Guineas love to fly and will roost in trees. They are not safe from predators here, but they are not as helpless as other fowl. You can train the guineas to use the shelter and nest boxes you provide if you start when they are young.

Quail

The most adorable eggs come from quail. A quail egg is about one-quarter the size of a chicken egg. You'd have to crack a lot of quail eggs to make a cake, but each bird lays multiple eggs per day. As meat birds, quail provide 6 to 14 ounces (170 to 397 g) of meat, depending on the breed. The meat is rich tasting, almost like a cross between chicken and duck.

They are small, grow quickly, and have a shorter egg production life than chickens. Eastern Bobwhite and

Cayuga ducks forage freely and lay dusky grey eggs.

Jumbo Bobwhite quail are sometimes found on the homestead, though the most popular for egg and meat production is the Japanese or Coturnix quail. These are the largest quail and the best egg layers.

Quail are meant to fly, and your farm is likely surrounded by wild quail, so you may find yours need a wire-mesh enclosure to keep them around. Give them 3 to 4 square feet (0.3 to 0.4 m²) per bird.

Quail are smaller and quieter than chickens and may be more readily accepted under urban and suburban ordinances.

Ducks

We usually have a dozen or more ducks on our farm. We raise our ducks for eggs, and they can be a good choice for meat as well.

Duck eggs have a much higher yolk-to-white ratio than chicken eggs, and the eggs are oilier, which makes them great for baking. Many people who are allergic to chicken eggs can eat duck eggs without problems. Ducks start laying eggs between 7 and 9 months of age.

Duck meat is rich and succulent. The fat rendered from duck is plentiful and a prized cooking fat. Ducks are much more difficult to de-feather than chickens and require some patience and skill. Depending on the breed, they may reach a good market weight of 4 pounds (2 kg) in 6 to 8 weeks.

Ducks are excellent foragers. We don't even have to feed ours until it gets cold and it's harder for them to find their own food on pasture. They need basic housing, with 4 square feet (0.4 m²) per duck, and produce more manure than chickens, which is something to consider when looking at stationary versus mobile housing. They don't roost like chickens and may be harder to coax inside at night. "Like a duck to water" is a common phrase because ducks really do love water and need it for their quality of life, even if it's as simple as a kiddie pool in their yard.

There are many breeds of ducks for the homestead. Indian Runner ducks make excellent layers, and it's a treat to watch them in the yard. With their upright posture, it looks like they're walking somewhere very important. Pekin ducks are good for meat production. Muscovy, Cayuga, Saxony, and Khaki Campbell ducks are good choices for both eggs and meat. (Muscovies are not genetically the same as other ducks; rather, they are their own species but still lumped into the duck category of poultry.) Duck math may become the next chicken math.

Geese

Goose breeds are classified as heavy, medium, and light by the American Poultry Association for exhibition standards. The types have little to do with their suitability for the farm. We keep Buff geese, a heavy type. Other popular breeds for small flocks are Embden, Toulouse, African, Chinese, and Pilgrim.

They are primarily raised for meat but also lay huge, rich eggs that are more like duck eggs than chicken eggs. Their laying season is short—just late winter through spring.

Our buff geese had a bad habit of sneaking into the garden and stealing tomatoes.

Heritage turkeys are much slower-growing than hybrid breeds, but they have a much stronger flavor and reproduce freely.

Geese are barnyard alarm systems as well, getting vocal when anything is out of the ordinary.

Geese can be ready to eat when they are 12 to 15 weeks old when supplemented with feed, and more like 16 to 24 weeks on pasture alone. Geese are even harder to dress than ducks. Their large size provides a good amount of meat for the effort.

Good pasture can provide most of a goose's feed requirements. If relying on pasture only, rather than supplementing with feed, plan for 1 acre (0.5 ha) of pasture for 20 to 40 birds. They are voracious eaters of seeds and can help get a weed problem in check. They rarely fly but do require predator-proof fencing. They like water and should have access, even if it's just a kiddie pool.

In a shelter, provide 3 to 6 square feet (0.3 to 0.5 m²) per goose. They do not roost, so dry bedding is especially important.

Turkeys
Turkeys are obvious homestead fowl. Their eggs are large, rich, and delicious, though turkey hens only lay for a few months in the spring, and they're not typically reliable egg layers at that. When you do get turkey eggs, you probably will want to keep them for hatching. I'll cover turkeys more in the next chapter.

CHAPTER

3

Meat in the Freezer

Raising animals for meat is the home-steading act most unfathomable to non-homesteaders. The idea of raising an animal and caring for its needs daily with the goal of butchering it is the most contested part of our lifestyle. People ask, "How can you care for something living knowing you're going to kill it?" Surprisingly, this question often comes from meat eaters.

The truth is, things die for us to live. Even in the production of vegetable crops, pests are killed to protect the plants. In the case of eating meat, even if you purchase it wrapped in plastic at the grocery store, it was once a living being. And unless that meat came from a pasture-based, sustainable farm, the life it lived was not a good one. It is because we care about animals that we take responsibility for our appetites and raise the meat ourselves.

Raising your own meat will create more gratitude for the meat you consume. In our house, it has eliminated a lot of waste. When you're paying $5 per pound for ground beef at the grocery store and you thaw it too soon and end up throwing it away, you might feel bad that you let that go to waste, but you would never let that happen with meat that you raised yourself. The value of it is so much higher.

"Were the walls of our meat industry to become transparent, literally or even figuratively, we would not long continue to raise, kill, and eat animals the way we do."

—Michael Pollan

Behind Industrial Meat

Before we began raising our own meat, Miah and I had already made the decision to source only ethically produced meat for our family. We supported small-scale, regenerative farmers, which was multiple times more expensive than the meat we were getting at the grocery store. Until you learn about the reality of industrial meat production and the true cost of raising meat responsibly, you can't fathom the price. Once you understand the differences between the production methods, you have no problem with the price. For us, the price difference meant eating less meat and being more thoughtful about the meat we were eating.

Before going any further, I want to emphasize that I respect and admire anyone who is producing food for themselves or others. Whether they're a quarter-acre (0.1 ha) homesteader or a 1,400-head beef cattle feedlot operator, this work is not easy; and the aim to feed people is noble. The reality for the animals and the environment, though, is that industrial meat production is not sustainable.

Concentrated animal feeding operations (CAFOs) are the feedlots and confinement barns where industrially raised beef cattle, dairy cattle, pigs, chickens, and turkeys are raised. Beef cattle usually spend the first half of their lives on pasture and the second half in the feedlot; pigs and poultry are born into and spend their lives in their barns. CAFO animals are not given the opportunity to be animals, see the sunshine, roll in the dirt, and forage. Their confinement systems produce the maximum amount of meat in the shortest period of time.

CAFO animals are fed medicated feeds to keep them from getting sick in their cramped quarters and to make them gain weight faster. This antibiotic overuse is the reason antibiotic-resistant bacteria are causing problems for human health. Antibiotic-resistant bacteria have been found in waterways and in the air we breathe around factory farms. The bacteria are also found contaminating food products.

The waste buildup is another problem. While we have septic tanks and sewage systems to manage human waste, these don't exist for CAFOs. They might dispose of animal waste by applying it to crop fields in the area, though improper application leaves nutrients to run off into waterways and aquifers. These nutrients are partly to blame for the dead zones in our lakes and oceans. CAFO waste might also be stored in pits or lagoons, which have the potential to leak and to overflow with heavy rainfall. You can imagine what all of this smells like. Emissions produce pollutants that often affect air quality in communities surrounding CAFOs and are a source of greenhouse gases, which contribute to global climate change.

On the flip side, not all animal agriculture looks like this. There is a beauty and a harmony to having animals on a homestead to nourish your soil and feed your family. Where CAFO farming concentrates waste and releases greenhouse gases, pasture-based animal agriculture has the potential to spread out nutrients and sequester carbon. Every time the ground is disturbed, carbon is released from the soil into the atmosphere. By keeping land in pasture for homestead livestock, carbon is transferred from the air to the soil, through the plants' leaves and roots, and held there. Increasing carbon storage in soils also improves water quality with reduced runoff, reduces soil erosion, and increases water conservation. These are reasons beyond feeding your family to keep land in productive pasture.

By giving animals the space they need to thrive, we don't need to give them low-level antibiotics just to survive. Their waste, too, doesn't build up and pose health problems, because we keep our animal numbers to a scale that won't overwhelm the land available to them. Sustainable homesteading means it's sustainable for the people, the environment, and the animals.

Black angus

Feed Sourcing

If there were a major feed shortage and the feed stores weren't able to supply our feed anymore, we would be putting a lot of meat in the freezer because we would not be able to sustainably feed all of our animals. Our goal is to feed our animals hay and grains grown on our own farm, yet we are not there.

It can be overwhelming to look at the ideal way of doing something and think you have to do it all the ideal way from the get-go. When we started raising animals, there was no way for us to afford the organic, non-GMO feeds that we wanted to feed. At the time, if I bought a chicken at the grocery store, I was buying a factory farmraised chicken for as little as $4. It does cost more to raise birds at home than it does to buy those factory-farmed chickens from the grocery store. Still, you are going to produce a healthier product and understand the care and expense that went into it, even if you purchase conventional feed, made with GMO corn and soy, to feed your own animals at home.

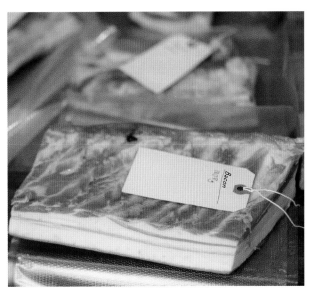

Cuts of pork are prepared for curing and smoking.

By raising our own meat, we feel the security of fully stocked freezers and the satisfaction of knowing no living creature had to live a miserable life to feed us.

Raising Poultry

As laying hens are the gateway to keeping animals on the homestead, broiler chickens are likely the gateway to raising your own meat, and turkeys are not far behind.

Poultry Breeds

Our first experience in butchering chickens on our farm came when we purchased Rhode Island Red chicks and well over half of them turned out to be roosters. We processed most of the roosters at 18 weeks and learned that heritage-breed poultry, like Rhode Island Reds, grow a lot slower and have a different quality and proportion of meat than breeds developed specifically for meat. This is true for chickens and for turkeys.

For a new homesteader, going with a breed specifically developed for meat production is the best way to get started. Cornish Crosses are the best-known hybrid meat chicken. We move them to new grass regularly and have had no negative experience with them, though with too much feed and not enough space, they are known to develop health issues. The Broad-Breasted White, which we've raised, and Broad-Breasted Bronze turkeys are the hybrid turkey go-tos. They were developed for industrial production but are suitable for pasture raising.

Reviewing the heritage-hybrid quandary from chapter 2, "Yard Birds," you'll remember that hybrid poultry have largely lost their mothering abilities. In the case of meat birds, the hybrids designed for industrial production cannot naturally breed. You have to buy each new batch of chicks from a hatchery. On the other hand, they are a fast-growing, economical source of meat for your homestead.

Hatcheries are always working on new homesteading meat chickens because there's a high demand. New breeds like Ginger Broilers, Red Rangers, and Black Rangers may be worth a look. They're only slightly slower growing than the Cornish Crosses, and they're developed to thrive on pasture.

If you want a more sustainable meat chicken—one that you can breed yourself on the farm—the heavier heritage breeds are your better option, even though they won't be as meaty or as fast growing as the hybrids.

Heritage turkeys are almost a different bird altogether. Their meat is darker and richer than the hybrid turkeys. If you've ever eaten wild turkey, think about that meat with a less gamey flavor. They're great foragers and mothers. Heritage turkeys have maintained their ability to fly, and you may find that one or more of yours will join their wild cousins in the woods. Because of their size, it's difficult to provide a fully enclosed pasture-ranging pen for heritage turkeys.

A few good homestead heritage turkey breeds are the Narragansett, Standard Bronze, Bourbon Red, and Slate.

Your chickens will be ready for processing in 6 to 16 weeks, depending on the breed and your management style. Heritage turkeys are grown out to market weight after 6 months of age, and hybrid turkeys after 3 or 4 months.

The Blue Slate turkey, Tom, adds character and beauty to the homestead.

Moving a flock of meat birds daily allows their waste to fertilize the pastures without killing the grass.

Poultry Housing

Turkeys and meat chickens, especially the Cornish Cross chickens, poop surprisingly more than laying hens. Chickens and turkeys raised outdoors require regular access to fresh pasture to keep them healthy and reduce the nutrient load on the land.

A chicken tractor shades the chickens from the sun, keeps them dry in the rain, provides a windbreak, and protects them from predators while allowing them access to the pasture and allowing their waste to fall directly to the ground. Chicken tractors are typical in pasture-raised poultry setups.

Chickens require 1.5 square feet (0.1 m²) each in the chicken tractor. If your chicken tractor is large enough,

it can hold your chickens 24 hours a day, or you can set up a temporary electric net fence around the chicken tractor to allow the birds to roam during the daytime.

Turkeys are larger and more boisterous than chickens, so predators rarely bother full-grown turkeys. Poults (young turkeys) are still at risk and need a secure enclosure. A structure that allows turkeys of all ages to take cover from aerial predators and the sun and rain is important, and perimeter fencing, such as the portable electric net fence, will help keep out potential ground predators and keep in your turkeys. Turkeys roost at night and will even fly to the roof of their shelter if they find that is their preferred roosting spot.

Breeding Poultry

Of the homesteading animals to breed, poultry may be the most hands-off. Your rooster or tom lives with your hens, and they take care of the timing of breeding. Your hens sit on the eggs and take care of the hatching and raising. You can incubate the eggs yourself in an incubator, which increases the hatching rate but also increases the attention and care required to breed poultry.

Turkeys lay eggs only in the spring, about one per day. You can eat turkey eggs, and they are rich and delicious, but to keep up your flock population, you may need to allow the hens to hatch what they lay. Turkey eggs hatch in 28 days, and the poults are especially fragile for the first 6 weeks.

Chickens lay more regularly year-round, slowing in the hottest months and when the sunlight dips below 14 hours a day. Chicken breeds meant for meat production will lay fewer eggs than those developed for egg production. Their eggs will hatch after being incubated for 20 to 21 days.

When purchasing day-old chicks and poults, be prepared to keep them warm and safe in a brooder area for several weeks before moving them onto the pasture. Without a mother hen to guide and care for them, these little poultry need time to develop their feathers and gain some size to brave the outdoors.

Lovely Cayuga duck eggs

ABOVE: A doe nearing birth has a nest box in her pen.

OPPOSITE: Rabbits being fed out for butchering are moved to new grass daily in mobile shelters.

Raising Rabbits

Many people have success raising rabbits as a homestead protein source for a number of reasons. Rabbits are great in a small space, even if you live in town. They are quiet, odorless, and not always governed by cities and homeowners associations. They are the fastest-growing animal and require the least amount of feed to produce meat. It's likely you won't need to spend more than a few hours per week caring for your rabbits, including feeding, watering, and moving their pens.

Housing Rabbits

Rabbits need all the protection from predators that they can get. Everything that can eat rabbits will eat rabbits. In the wild, rabbits are fast and agile, but in a pen, they have fewer means of escape.

Offer rabbits a sturdy enclosure, whether mobile or stationary. In the summer, it's especially important to provide lots of shade; in the winter, rabbits need to get out of the rain and wind. Mobile rabbit housing operates like chicken tractors. They're essentially hutches on wheels surrounded by or attached to a fenced enclosure. Stationary housing can be cages hanging from the rafters of a shed, stackable cages, a stand-alone hutch, or a whole rabbit section in your barn. Consider how much space you can allow the rabbits to roam and how much time you want to spend moving them from hutch to pasture.

A rabbit doe about to give birth needs a nest box. Given the right materials in her enclosure, she may build her own nest, or you can provide her with a bed of straw or grass clippings for her newborns. Kits (baby rabbits) are born without fur and need this extra protection. Does and kits should share an enclosure until kits are weaned, around 4 weeks of age.

Rabbits will forage grass and forbs, and they'll need a feed supplement, too. Commercial pellets make the most sense for complete nutrition. In small amounts, fresh feeds—like root vegetables, cabbage leaves, and grass clippings—help stimulate their appetites and maintain fiber in their diets, but this shouldn't make up the bulk of their feed.

You may be surprised by the amount of water a small animal like a rabbit drinks. A doe and a litter of kits will go through 1 gallon (4 L) or more each day.

Rabbit manure can be collected and put directly on a garden without the risk of burning plants.

Rabbit Waste

Rabbit poop is a rich garden benefit. It's about 2 percent nitrogen, 1 percent phosphorus, and 1 percent potassium, so it's a well-balanced amendment for your compost pile. It's also not offensive smelling like other livestock waste.

Another use for rabbit manure is to grow worms for fishing or vermicomposting. Worms love rabbit poop.

Breeding Rabbits

Like chickens, it's easy to end up with more rabbits than you need. As I have emphasized throughout this book, return to your goals in building your rabbit population. Depending on the breed, rabbits can grow to 4.5 pounds (2 kg), a good meat weight, in just 10 weeks. If you want to grow two fryers for your family to eat each week all year long, you only need four breeding does and one breeding buck (male rabbit).

Rabbits are a great food source because of how quickly they reproduce and grow.

Rabbits reproduce rapidly: Gestation is only 28 to 31 days. One doe can have 6 to 8 litters each year with an average of 5 kits and as many as 12 kits per litter.

Rabbits will breed year-round, though may be slower to breed when it's very hot and humid. Assuming the doe is healthy and in good body condition, you should breed her again 35 days after she's given birth. A regular schedule keeps your doe in her best breeding condition. We found that managing this schedule was one of the most challenging parts of keeping rabbits.

Rabbit Breeds

You'll find more than 50 breeds of rabbits for show, fiber, and meat. While the show and fiber breeds are beautiful—some with tiny ears, some with shaggy coats—they aren't as useful for raising meat fryers. New Zealand White is a good meat rabbit breed. Crossed with Californians or Champagne d'Argent, their meat yield is even better. I like American Chinchilla and Silver Fox as well.

Find your rabbit breeding stock from someone who produces meat rabbits rather than the flea market or pet store.

When regularly handled, breeding rabbits can be very easy to manage.

ABOVE: Mangalitsa pigs are commonly identified by their wiry, curly hair.

OPPOSITE: Mangalitsas can be kept in warmer climates, but they need access to fresh water and a place to wallow in the mud.

Raising Pigs

You probably don't need a large team of pigs to produce the amount of pork your family wants for the year. We keep just two breeding sows and one boar for the eight of us. Pigs are pregnant for 3 months, 3 weeks, and 3 days. They have multiple piglets per litter, so your pig population and pork supply can expand rapidly. Depending on the breed and your management, you can expect 160 to 200 pounds (72.5 to 90.5 kg) of meat from each pig processed.

Pastured Pigs

Unlike that of other animals on the farm, pig manure doesn't make great compost. It's lower in nutrients, carries more parasites and bacteria, and keeps that pig poop smell, even after composting or being incorporated into the soil. After letting the pigs work up an area, I would not plant anything that grows touching the ground the next season for food safety.

In nature, the pig's nose was designed to dig. Given a new space, they'll first get to work creating a hole to wallow in, as mud wallowing cools them off. Pigs till the ground and can tear it up, which may be good or bad, depending on your objective. The pigs did do a good job of unearthing rocks for us to remove on our first farm, which had such rocky soil. Pigs can clean up the garden after it's been productive, yet they can also compact the soil in the process, so plan their garden duties for a dry period and pay attention to how they're treating the land.

A well-established pasture will support pigs if they're moved often enough to maintain most of the vegetation on the land. Pigs also thrive in forest spaces. Trees' shade keeps them cool, their rooting can remove invasive underbrush, and they can forage on nuts and berries.

Mothering instincts in a breeding sow are highly valued.

Feeding Pigs

Your meat pigs and breeding pigs have different nutritional needs. Keep them separate for ease of feeding. You want your feeder pigs to grow quickly and put on more fat, whereas you want your breeding pigs to be lean.

Foraging in the pasture or woodlot could reduce pigs' need for supplemental feed by a third to a half. Pigs need a lot of protein that foraging likely doesn't offer. A complete feed from the feed store will take care of the difference.

In the winter, pigs will eat more feed because there's less to forage and they're trying to stay warm. Try a deep-bedding system in their shelters. It gives off heat as the waste and bedding break down, which helps reduce their feed needs.

Pigs drink a lot of water: 6 to 8 gallons (22.5 to 30.5 L) each per day in normal weather and as much as 12 gallons (45.5 L) in hot weather. They can be trained to use an automatic waterer attachment on the end of a hose, or you'll have to provide a trough low enough for them to access. Be aware that if your pigs can knock over their water tank, they will.

Fencing and Housing Pigs

Pigs require minimal housing—just something to provide shelter from sun and wind. Miah cut a hole in the side of a few food-safe plastic intermediate bulk container totes as simple, portable pig houses. The totes were free from a local company that had no use for them. A three-sided shed, a straw bale enclosure, or another repurposed shelter that can be anchored against the weather would do.

There's that saying that fencing needs to be hog tight, horse high, and bull strong. Pigs have an impressive ability to find a fence's weak spot and break out. They don't need super high fences—3 feet (1 m) is good—as they typically don't climb or jump. They can dig, so the bottoms of fences should be fortified with electric wire; piglets need bottom wires just a few inches apart so they don't squeeze out. High-tensile fence and hog-wire fence work as a permanent perimeter fence. Electric wire or electric net fencing is easy to move to make smaller pastures for rotation.

Pig Breeds

Not any pig will thrive in a pasture-based system, as modern breeds have been developed for confinement-feeding operations. Heritage breeds are a good option because they were developed to thrive outdoors. They generally carry more fat and meat flavor, too. Breeds developed for industrial production gain weight faster but may not forage as much or be as healthy.

Popular pastured-pig breeds in the United States include Yorkshires (the heritage type, not the commercial breed), Large Blacks, Gloucestershire Old Spots, Berkshires, Tamworths, and Hampshires. We raise Mangalitsas, which are best known for their wiry hair. They were nearly extinct in the 1990s, and now their numbers are growing in the United States. Their meat is sometimes called the Kobe beef of pork because of its marbling and high-quality fat texture. Mangalitsas tend to have smaller litters than other breeds, which is something Miah was looking for because he didn't want to get overwhelmed with our pig population.

TOP: One-day-old Mangalitsa piglets
ABOVE: Pigs make quick work of any scraps from the garden.

Breeding Pigs

To be sure pigs are right for you without investing in breeding animals, you may start with a few feeder pigs to raise and process first. When it is time to move into pig breeding, know that pigs don't generally need help with farrowing. Start your pig-breeding experience with a sow who's had a litter so at least one of you knows what you're doing and try to have a pig mentor on hand for your first farrowing to talk you through what's happening.

We keep a boar, but you could transport your sow to a boar or use artificial insemination, depending on the breed you're keeping.

Belted Galloway

Raising Beef

Bringing beef cattle to the farm takes some consideration. Even the smallest breeds of beef cattle are big. They are strong, and they consume a lot of forage. They also provide a lot of food for your family.

Cattle really don't like being alone. While your family may only need the meat from one beef animal, raising multiples is important for their sanity and yours.

Feeding and Processing

As a ruminant, a cow was not designed to digest grains. The feedlot industry began feeding grain to fatten beef animals faster, and this has led American palates to prefer the taste of grain-fed beef over grass-fed beef.

It's your choice to raise your beef animals with a mix of forage and grain or feed forage only for truly grass-fed beef. A 1,100-pound (500 kg) cow will consume 22 to 33 pounds (10 to 15 kg) of forage or half a square bale of grass hay each day. There is a science to high-quality grass-fed beef, which includes intensive rotational grazing and careful breeding. Your grass-fed beef cattle can take about twice as long to reach market weight as grain-fed cattle: about 36 months versus 18 months.

You also have the option of processing your cattle as veal, when they weigh 700 pounds (317.5 kg) or less. Veal offers much less meat, but you don't have to feed them through the winter, and maybe your family doesn't need that much meat anyway. A 1,200-pound (544.5 kg) beef steer can return 490 pounds (222.5 kg) of beef, whereas a veal calf will provide less than half that.

Breeding Cattle

Bulls are a whole topic. They have minds of their own, especially when cows they're interested in are nearby. To maintain a homestead beef herd, you're better off sending your cows to a bull on another farm for breeding. You'll need a trailer for transporting them, but it will save you the trouble and expense of keeping a bull whose services you need just once a year. Artificial insemination is another breeding option. This is big business in commercial cattle operations, but you can't typically find heritage-breed genetics in artificial insemination.

To start your beef-raising experience, consider whether you need to breed your cattle at all. You may begin by purchasing weaned calves for the first few years so you can learn how to manage cattle and find the right breed. When it's clear that raising beef cattle is right for you, bringing in breeding stock is a good next step.

Cattle have a 283-day gestation period, plus or minus a few days. Each cow typically has one calf but sometimes two. The average beef cow can be rebred in 2 months.

Hereford

ABOVE: Beef cattle do not need a fancy barn. We use our barn for milking and storage. Beef cattle will rarely even be in this barn.

OPPOSITE: Breeding a dairy cow to a meat breed bull will result in a calf that can be dual purpose. This Jersey/Red Poll cross heifer could be raised as a family milker or raised for the freezer.

Housing and Fencing

You may have heard that it's easier to fence in cattle than it is to fence in chickens. This is true. It seems backwards because cattle are big and chickens are not, but cattle can be trained to respect a few strands of electric fence in a way chickens can't. As long as your property has a strong perimeter fence in case something were to go wrong, your cattle can exist behind two strands of electric wire fence.

Beef cattle also do not require extensive housing. They need a source of shade in the summer and a wind block in the winter. If you did provide them with something fancier, they probably wouldn't even use it.

Because of their size, working with your cows requires some thinking. A corral with a narrow alley and a squeeze chute to restrain them for the times you need to handle them, like during health issues and pregnancy checking, minimize confusion and stress for you and your animals.

Cattle Breeds

There are many excellent heritage breeds of cattle to choose from. While I usually don't recommend dual-purpose breeds, cattle are the exception. It takes a lot to breed cattle for either meat or milk (more on the home dairy in chapter 4), and if you can find one that does a good job of both, it may be worth your while.

The most popular beef cattle breeds in the United States are Black Angus, Red Angus, and Herefords, and cattle breeders have begun developing lines for the pasture-based farmer. If you're looking toward these breeds, be sure to get one that's right for your system. Cattle whose genetics are meant for a feedlot will not thrive on the small farm.

Pineywoods is a rare dual-purpose breed that is one of the original cattle breeds brought to the United States. It's mostly used for meat production but can be used for dairy purposes as well. Devon, Belted Galloway, and Highland do great on pasture in small-farm settings.

Miniature beef cattle are an option if you have small acreage: Lowline Angus, Miniature Hereford, Miniature Zebu, and Dexter among them.

Raising your own meat animals brings your homesteading work full circle. Acquiring animals is the easiest and least expensive part of this. Harvesting your animals may be the hardest part. It will teach you gratitude and humility that you didn't know was possible.

Before you go bringing home every cheap pig and good deal on a turkey to cross your path, stay rooted in what you know is right for you and your homestead. If you don't have a place to put it, don't have a way to feed it, or don't have a plan for the meat in the end, this animal is not a good deal. Planning and patience will help you do things in the right order, when you're physically and mentally ready to do them. Another cheap pig will come your way next season, I can almost guarantee.

ABOVE: A dual purpose Jersey/Devon cross heifer

OPPOSITE: Involving children in the process of raising meat and taking responsibility for our appetites gives them a greater respect for the cost of our food.

TOP: *Butchering meat birds is faster with a group.*

CENTER: *Proper handling is crucial when processing meat at home.*

ABOVE: *Doing your own butchering is more work but provides the freedom to create exactly the product you want.*

Veterinary Care

Keeping animals on the farm means having to be honest about what you can afford to spend to maintain their health. High-quality feed is not cheap, and veterinary bills are even less so.

Giving injections, drawing blood, and assisting in births are a few of the procedures you'll quickly learn from your veterinarian and your experienced friends. With each new challenge, you'll become more confident in working with your animals and learn more about their care.

We made our first vet call when Doris, one of our Mangalitsa pigs, birthed her first litter. We were concerned she was having a prolapse, and, after consulting farmer friends by phone, we decided we needed a veterinarian's help. It's hard to admit that not every animal emergency will warrant a vet call. Some will lead to learning how to do things on your own, and some will result in needing to cull the animal. Raising animals leads to hard decisions, and we have a cap for what we're able to spend on any animal on our farm.

Butchering at Home vs. Outsourcing

There is a recognition and a gratitude in harvesting your own meat. Miah does the processing, and he doesn't take it lightly. It doesn't seem to get easier. True sustainability means processing your own animals yourself, yet this is not for everyone, for multiple reasons.

I think it goes without saying that the smaller the animal, the easier it is to process. A small group of people can make quick work of processing a couple dozen chickens or a few rabbits. Turkeys may take a little longer yet are easily doable at home. Pigs and cattle require more equipment and skill.

If you have never taken part in processing animals, there are videos out there that can teach you, but there is no substitute for hands-on learning. The larger the animal, the more important your confidence becomes. Skilled butchers will not only treat your animals with respect at harvest, but they will also maximize the amount of usable meat from them during the cutting process.

There's no way to get into the details of home processing here, but here are a few basics:

• Know the rules regarding animal processing and disposal of offal—meaning skin, bones, blood, and organs—in your area.

• Ready your tools. For large animals, this includes a stunning device, such as a .22 caliber rifle; a chain hoist to secure to a barn rafter or a tractor equipped with a hydraulic lift; and a meat saw. For poultry, you need killing cones to hang the birds upside down; a scalding tank; some means of plucking, whether that's a plucker machine or ambitious help, and pinning knives; and an ice bath for quick cooling. To process any animals, you need sharp knives in various sizes, sturdy worktables, and containers for holding the offal.

• Choose a location that's secure from pets and rodents. It should be dry, clean, and dust-free with no odors. It needs running water, even if just from a food-safe hose. Cleanliness is essential for your processing equipment, hands, clothing, work area, and storage area.

• Slaughter on a cool day. Meat needs to be cooled quickly for food safety, and beef needs to hang and age for 3 to 7 days before butchering into cuts. Night temperatures should be 32°F (0°C) or lower for the carcass to chill properly without refrigeration.

If you decide to outsource your processing, you'll find a supply and demand strain. In some places, butchers are booked a year or more in advance. If you're breeding rabbits, buying feeder pigs, or starting with hybrid poultry, you may need to set your processing dates before you even get your animals.

Skilled processing is an investment. If you're eating this meat at home and not planning to sell it, save money by going to a custom-exempt slaughterhouse. These cost less because they don't have to pay a federal meat inspector. If you are selling your meat, you may need USDA inspection, depending on your state laws.

Processors generally charge by the pound for slaughtering large animals and by the animal for rabbits and poultry. Save money by bringing home primal cuts of beef and pork and breaking them down into steaks, roasts, and ground meats yourself. Here, again, you'll get

TOP: *The offal and caul fat of a hog.*

ABOVE: *Sharpening knives by whetstone ensures an easier time butchering. Sharp knives are imperative for safety.*

the highest-quality product from someone with experience, but it is something you can learn.

Some areas also have poultry and small-animal mobile processing units, often hosted by a university or sustainable-farming group. This hybrid approach allows you to process your own animals without needing to invest in the facilities and equipment.

THE FIRST-TIME HOMESTEADER

HOCK

—04-03-2022—

Weight: 1030

2.5% Salt 26

1% Sugar 10

0.5% Black Pepper 5

ABOVE: Properly cured meat can last a very long time without refrigeration.

OPPOSITE: Our friends, Andy and Doug, from Hand Hewn Farm in Ohio, taught us to properly measure salt to cure pork before smoking.

Storing Meat

An often-overlooked cost of raising animals for meat is where to put that meat come processing time. You're going to need a lot of freezer space. If you raised just one beef steer, one pig, and 30 broiler chickens, you need 30 cubic feet of freezer space: about 1 cubic foot for every 30 pounds (13.5 kg) of meat.

Freezing slows meat degradation but doesn't totally stop it. Long-term storage requires 0°F (−18°C) or below. Chicken and turkey kept at this temperature should maintain its quality for a year; warmer than 0°F (−18°C) and it may only last for 3 months before you notice a difference in quality. Ground products have a shorter ideal freezer life (up to 3 months), while roasts, chops, and steaks are generally good for 6 months to a year.

Correctly packaging meat is key to retaining quality. Use moisture-proof packaging to reduce freezer burn, which affects the quality of the meat but generally not the safety. Shrink wrap, plastic freezer bags, butcher paper sealed with freezer tape, and aluminum foil are good choices. Write the date, weight, and cut of meat on each package because you really won't remember what you put in there 6 months later.

CHAPTER

4

The Home Dairy

There's something terribly romantic about hearing the screen door slam behind you as you walk out of the house to the milk stall every morning, milk bucket in hand. Within this twice-daily ritual, your dairy animals almost become a part of your family. You commune, getting to know their bodies as well as you know your own. They become the farm's timekeepers. Your mornings and evenings are punctuated by a visit with your lovely ladies.

The rhythm of a home dairy is beautiful, and it is hard. Milking happens twice a day, every day, whether it's raining, snowing, or blazing hot. Your cows and goats need you, even when you're looking forward to a family wedding, child's dance recital, or night on the town. The milk doesn't stop flowing on holidays, when you have the flu, or when you need a vacation.

If your family consumes dairy products, keeping dairy animals helps offset the need for reliance on the grocery store. They provide milk for drinking, cooking, and making cheese, yogurt, kefir, and more. There is a lot of work involved in keeping them, and there is a lot of reward, too.

"As for butter versus margarine, I trust cows more than chemists."

—Joan Dye Gussow

ABOVE: *Most goats, unless they have polled genetics, will grow horns. This is avoided by disbudding (having their horn buds removed) before the goat turns two years old.*

OPPOSITE: *Our Jersey cow, Hope, is the queen of our homestead.*

A Few Goats or the Family Cow

When you have a dairy cow and you're getting several gallons of milk a day, it might feel pointless to have goat, but I cannot imagine not having goats. I love their personalities. At the same time, when Hope, our first cow, came to the farm, a new piece of my heart sprung up and overflowed with her arrival. She sends 20 pounds (9 kg) of fresh, raw milk through my front door every morning. Her service to us is humbling, and I am in absolute admiration of her.

I will be no help in your decision between having goats and cows if you need to choose between the two, but I can offer some facts about them.

Cows are much larger than goats. As a result, they need more space, they eat more feed, and they command more care in handling. Cows also cost more to purchase, yet they are easier to fence in and keep out of trouble than goats. Goats tend to be better suited for smaller properties and ones with less top-quality forage, they're

Finding a friend that can teach you how to milk—hands on—will prepare you for having a dairy animal of your own. Here, my friend, Justin Rhodes shows me how to milk a cow for the first time.

Gallons of fresh milk every day leads to a whole new level of food security.

less expensive to get started with, and they're good for homesteading families whose kids want to be involved in milking.

Cows produce more milk per day for a longer period after giving birth, but goats have a longer productive life. Your cows can milk about 300 days per year, producing 3 to 9 gallons (11.5 to 34 L) per day, depending on the cow and your management. You can generally milk cows for 4 to 6 years. Goats are in milk for about 284 days per year, producing 2 to 4 quarts (2 to 4 L) per day. They should produce milk for 8 to 10 years.

Consider, also, the end product. Goat milk and cow milk taste different. Goat milk has a more grassy, earthy flavor than cow milk. Both cow milk and goat milk have lactose, a sugar that some people have trouble digesting, but goat milk's smaller fat globules may make it easier to digest. Goat milk is higher in calcium, vitamin B_6, potassium, and niacin, while cow milk has more B_{12} and folate. Goat milk is better suited to making soft cheeses and cow milk to a wider range of cheeses.

Home Dairy Beginnings

It makes sense to think your home dairying begins when you get your first dairy animals. This is not actually the place to start. Mentorship is where your dairy begins. Finding someone to share hands-on experience with you is monumental.

It's helpful if you have experience caring for farm animals to begin with, yet milking and caring for dairy animals adds another layer. Aside from giving you the hands-on experience you need to confidently handle dairy animals on your own, mentorship helps you refine your dairy dreams. You might learn that you love cows but have no patience for goats, that you prefer hand milking over machine milking, or that having dairy animals isn't your thing after all.

When it's time to bring home your first dairy animals, start with those that have some experience. An experienced dairy animal can mentor you nearly as well as an experienced farmer. Animals new to milking need to be trained with some level of confidence.

Milk genetics, too, are important in selecting your first animals, and these vary greatly within the breeds. Experienced breeders heavily consider good milking traits when selecting their animals. Keep in mind that production isn't solely about the genetics. A consistent milking routine and well-rounded diet contribute to maximum production.

When looking for your first goat or cow, look to reputable farmers with disease-free herds. Find out how much milk the goat or cow you want is giving, ask about her health history, and watch her being milked.

Dairy animals must be milked every day, no matter what. This makes a home dairy animal a huge commitment, but it is one worth making!

From left to right, a Nubian doe, Lamancha doe, and Nigerian Dwarf doe.

Hand milking a couple of goats can be faster than the time required to clean a milking machine.

Setting Up Your Dairy

The second step in your home dairying experience is to prepare your farm. This is another place where I'll admit that we got ahead of ourselves in bringing home our first cows, Hope and Honey, before we were quite ready for them. We were close to being ready, thanks to years of having dairy goats and because we were in the process of building a barn.

Before your animals come home, you need proper fencing and housing, a milk room, and a handful of equipment and supplies.

Fencing and Pasture

The pasture space you have available is a driving force in whether you choose goats or cows for your home dairy.

Managing your pasture properly is smart for the health of your animals and land, as well as for your feed bill. You could turn out your animals into a big field and let them graze as they please, but to get the most from the forage in the pasture, to allow the most valuable plants to regrow, and to keep your animals moving away from their own waste, rotational grazing is the way to go.

The amount of pasture space needed for each herd depends on many factors. How many animals do you have? How many are lactating and pregnant? What type of plants are in your field? What time of year is it? Are you in drought? Proper pasture rotation requires that you pay attention to the condition of your animals and the condition of your field and to adjust from there.

Building fencing to keep the goats contained can be a challenge. If a goat is able to climb or get its head stuck in a fence, it'll find a way to do it. High-tensile fence, woven wire made for goats and sheep, and electrified net fencing at least 48 inches (1.2 m) high are a good start.

Unless there's something on the other side of the fence that they want, dairy cows probably aren't going to test their fence. Keep them watered, fed, and milked and they'll go easier on the infrastructure than their goat friends. We use a woven wire perimeter fence, electric fence within it to further break up the pasture, and corral panels to make smaller paddocks. They will also do fine with electrified net fencing or a few strands of electric wire.

ABOVE: Gabriel, our Great Pyrenees, is a livestock guardian dog. He lives with the animals and deters predators with his threatening bark.

BELOW: Our herd of dairy goats consists of multiple breeds. They are heavy bred in late winter, just a few weeks shy of kidding time.

Housing

A free-choice three-sided shelter or other means of getting them out of the sun, wind, mud, and ice are all your dairy animals need on a regular basis. As people, we like to be comfortable; and you may want to give them a fully enclosed, heated space for cold weather. This, however, will cause more health issues than not. In the shelter, give each cow 50 square feet (4.5 m²) of space and each goat 10 square feet (1 m²).

Come birthing time, a secure birthing area makes life better for you and your animals. These could be permanent stalls or something you build with portable panels when needed. Pens for birthing calves should be 150 to 200 square feet (14 to 18.5 m²); kidding pens should be 20 square feet (2 m²) or more.

Bed the birthing stall well; straw is ideal. A three-sided shed is adequate for birthing, though an enclosed barn adds a nice comfort element. Cold and wet is the worst combination for bringing baby farm animals into the world.

Calf-sharing means milking a cow while still allowing her to raise her calf. This means less milk for the farmer, but far more flexibility and, often, healthier calves.

Hand milking is an intimate and cathartic act. It is, without contest, my favorite farm chore.

The Milking Room

Your dairy girls need to be comfortable to produce their best and stay healthy. A space on your farm dedicated to milking gives you one place to keep your milking equipment, allows you to keep a clean area, and gives your animals a routine. They know when and where to meet you for milking.

Our first milking room was a space we added on to the side of the goat barn. We sided it with pallets and outfitted it with crates hanging from the walls to store supplies. It provided shade, protection from the wind and rain, and a safe, consistent place to milk. When we first got our cows, we were milking Hope in the field. This is how she was milked on the farm she came from, and this helped to ease her transition. We soon built a stanchion for milking cows at the barn.

Your milking room needs a milking stand or stanchion, which simply consists of a bucket for feeding, a head gate to keep the animal stationary, and a nonslip platform. You may also install side rails for your cows. A goat-milking

platform should be 15 to 18 inches (38 to 46 cm) higher than the floor to save your back. If you have a small cow breed, they may also need a raised platform; full-sized cows can usually be comfortably milked from ground level.

We hand milked until we got our second dairy cow. We held off on getting a milk machine because I love hand milking and I hate cleaning. If you're milking a single cow or a couple of goats, it takes less time to milk them by hand than it does to clean and sanitize a milk machine and all its parts. Some people choose to milk by machine because it's completely closed, which means there's less chance of contamination.

Most of your milking supplies can stay in the milking room. I carry the buckets back and forth from the house because I pour off milk in the house and need to sanitize the buckets. Everything else—cleaning supplies, feed, teat dip, clean towels—stays in the milk room.

Milking Time

This preparation eventually leads to the reason you're doing this work: the freshest, creamiest, most nutritious milk you've ever experienced.

Come milking time, your girls will be waiting for you. For one, they look forward to their supplemental grain, which they eat while being milked. And two, after 12 hours of holding their milk, they're ready for you to relieve them of it.

You'll learn the quirks of each of your animals. Mayhem, one of our first dairy goats, would give me the hardest time at milking until I figured out she wanted me to milk her from between her hind legs rather than from her left side. Listen to your animals and they will cooperate with you.

Let each cow or goat into the milking room one at a time. More than one turns into chaos. Put feed in the bucket and she'll put herself onto the milking stand. Secure the head gate and get to work.

I cannot emphasize enough the importance of cleanliness. Using a rag in warm water with antibacterial soap, I wash off their udders before I start milking to reduce the hair and dirt that might come off the teat or udder into the milk bucket. This also washes my hands.

The process of milking is easier experienced than explained, but you'll learn the rhythm and won't even have to think about it after your third try. In short, you massage the udder a bit to signal her body to let down

Nigerian Dwarf goats often have beautiful blue eyes.

THE FIRST-TIME HOMESTEADER

Milk expressed by hand must be strained through a filter or cloth to remove any debris that may have fallen in during the milking process.

milk. When milk fills the teat, you squeeze the top of the teat to hold the milk in place and then squeeze from top to bottom to release the milk.

Use the first few squirts of milk to strip the teats of anything that's in the milk duct. Direct this into a small bowl or cup, not into the milk bucket, to examine it for weird chunks or blood, which denote infection. Assuming it's all good, which it will be most of the time, milk into the bucket until you feel there's no more milk. It's important to milk out as much as possible for health and optimum milk production.

After milking, give the udders and teats another wash with the soapy water to be sure you're not introducing infection. In cold weather, I like to apply a little olive oil for moisturizing. Let her finish eating her grain and use this time to check her over, looking at her body condition, hooves, and any bumps or scratches.

When everyone is finished milking, clean off the milk stand using the bucket of wash water.

If, during milking, the cow or goat were to stick her hoof in the bucket, which sometimes happens, or if there were any chance that poop or another contaminant got into it, don't drink that. Feed it to the pigs. While Hope and I were getting used to one another, I milked her to about halfway into one bucket and then switched to a fresh bucket, so if something happened, like she knocked over or stuck her foot in the bucket, I wouldn't lose all the milk—just half of it.

Do not delay in putting away your milk. Get it really cold, really fast for flavor and food safety: Chill to below 40°F (4°C) within 2 hours to stunt bacterial growth. Filter the milk through a milk strainer into sanitized jars, label the jars with the date and type of milk, and put them directly into the freezer. I set an alarm on my phone for an hour so I don't forget to move the jars from the freezer to the fridge. Only do this with freezer-safe jars, of course, and leave ample headspace in glass jars so you don't end up with broken glass and milk all over your kitchen, even if you do accidentally leave a jar in the freezer.

Goats that have been handled and socialized prove to have lovely personalities. They make great pets and can bond with people the way you would expect a dog to bond.

Feeding Your Herds

A milking animal eats about 3 percent of her body weight in hay or pasture every day. That means a large cow, like a Holstein, eats more than a smaller cow, like a Jersey, and a goat eats even less.

An animal that's fed well produces more milk than one fed poorly. Goats and cows are ruminants, meaning they have a four-part stomach designed to digest forage, such as pasture and hay. This should make up most of their diet. Goats are efficient browsers and will get what nutrition they can from weeds and shrubs. Cows have less ability to digest plants other than grasses and legumes.

It is possible for your dairy animals to get all necessary nutrition from pasture, but it's difficult. They won't produce the volume of milk on grass alone that they will with grass and supplemental feed. We supplement with feed because, without it, we'd need to source some of our dairy needs from store-bought dairy, which is almost certainly from cows fed supplemental grain.

Our supplemental feed is a combination of several types, including complete pellets, sweet feed, and alfalfa pellets. We use this mix because occasionally the feed store will run out of a certain feed. If one component is missing, this doesn't drastically change their overall feed, which could cause digestive and production problems.

A doe needs about 1 pound (453.5 g) of supplemental feed for every quart of milk (1 L) she produces. A cow needs about 1 pound (453.5 g) of feed for every half-gallon (2 L) of milk.

Water access, too, is vital. Milk is made of 87 percent water. Cows need at least 4 gallons (15 L) of water to produce 1 gallon (4 L) of milk; goats need a half-gallon (2 L) for every gallon (4 L) of milk. Cows will drink 30 to 50 gallons (114 to 189 L) of water per day and goats up to 3.5 gallons (13 L) per day.

Bulls, Bucks, and Breeding

For an animal to produce milk, it must first be bred and give birth. Unlike raising animals for meat, breeding is a required part of the home dairy.

We have always kept bucks on the farm, and we brought home a bull shortly after getting our cows, too. We feel it is the more sustainable route to have direct access to a male for breeding. Keeping bulls and bucks on the farm requires a separate pen—they can't even share a fence line with the females of their desire—and an added measure of respect when handling. Diversity in your herd is also something to consider. You may need to sell and buy a breeding male every few years to keep your herd's genetics fresh.

Many homesteaders choose to transport their dairy animals to a stud service, whether that's on a nearby farm or to a breeder with genetics they want. Others opt for artificial insemination (AI). With these two off-farm options, you can access highly rated stock with a lower investment because you don't have to house, feed, and care for the male animals throughout the year. Using AI, you may even be able to source semen that has been sexed, allowing you to stack the odds for female offspring, which are more valuable than males.

In breeding for a dairy, you'll end up with more animals than you can support, and it's easy to get attached to them. At some point, you'll need to sell or trade some or to raise them for meat.

Two-week-old goat kids healing from disbudding. There are valid arguments on both sides of the horns/no-horns argument, but if you plan on showing any goats in competitions they will have to be disbudded.

Birthing and Baby Care

Looking at a ruminant's abdomen, left is lunch, right is baby: Their stomachs sit on the left side of their belly, and their babies are carried on the right side. When you can get up next to them and feel the baby kick, reality sets in: You're really doing it.

Understanding a little physiology goes a long way in determining your breeding schedule. Ideally, you'll stagger breeding so you have at least one animal in milk all the time.

For goats, gestation (pregnancy) is 145 to 155 days. They are seasonal breeders, coming into heat in late summer through early winter, according to the decreasing length of daylight. You can trick your goats into heat by putting them in a dark barn for part of the day to simulate decreasing daylight at other times of the year.

Doelings (female baby goats) can be bred when they're 7 to 10 months old, or 80 pounds (36 kg). Nigerian Dwarfs, which only reach 75 pounds (34 kg) at maturity,

should be at least 50 pounds (23 kg). Bucklings (male baby goats) are fertile by 5 months of age, so it's important to separate them from the females by this time to prevent accidental breeding.

For cows, gestation is 279 to 287 days. A cow can be bred back 55 or so days after calving, depending on her age, body condition, and how the birth went.

A heifer (a female cow that hasn't given birth) can be bred at 12 to 14 months old. Bull calves can begin breeding females at around 12 months.

Expect your doe to have one or two kids, though you may end up with three. It's rare for cows to have multiple calves. When a cow has twins, if one is male and one is female, the female is called a freemartin and is almost always sterile. This isn't the case with goats. They can carry boy-girl multiples with no repercussions to the kids' sterility.

Nubian triplet kids at one day old

Cows have a gestation period of 279–287 days.

The Birthing Kit

Put your supplies together in one place ahead of birthing time. It's easy to grab a box or a bag and head to the barn at a moment's notice. It's harder to hunt around the farm for the items you swore you saw somewhere.

Check your birthing kit as that time approaches to be sure it has all you need in stock. I suggest the following:

• Syringes with needles

• Bottles for feeding. You can buy milk bottles specifically for this, but we use 16-ounce (473 ml) plastic soda or water bottles.

• Nipples for bottle feeders. We use Pritchard nipples that screw on to the plastic bottles.

• Iodine for dipping umbilical cords

• Milk replacer and colostrum powder (unless you have a goat or cow already in milk who you can borrow from and colostrum already frozen)

• Blackstrap molasses to mix with water to give the doe or cow a boost after kidding

• Clean towels

• Alfalfa pellets or other high-nutrient feed to offer after birthing

• Dewormer to administer to does within 24 hours of kidding because their immune system is compromised after giving birth and their parasite load can get high

• Thermometer

• Flashlight

• Rubber gloves

• A scale if you want to weigh your kids

• A fan when it's hot and flies are a problem

• A heat source, such as a heat lamp or warm towels, when it's cold

Birthing

According to The Doe Code—the code by which all does (and cows, too) subscribe—an animal will give birth at the worst time possible. Miah and I have missed dinner dates, I have missed social outings, and just forget about normal sleep around birthing time. They also seem to wait for the coldest night or the biggest storm to want to have their

A properly outfitted birthing kit is essential to ensure you have everything you need when babies like this one arrive.

babies. You can read everything about birthing and still not feel confident until you experience it yourself.

When you think birthing is near, be sure the birthing stall is clean and well bedded, and bring in the mother-to-be. Particularly when the weather is below freezing, be as attentive as possible to animals about to give birth.

The momma will start to show discharge close to birthing time. Her udder will fill, and the ligaments around her tail will get soft and nearly sink away. As labor begins, she'll start to act restless.

You shouldn't need to assist, but there's always a chance that a baby is very big or presenting in the wrong position. It's hard to watch what might look like a struggle, but you should generally wait 30 minutes before intervening.

Immediately after birth, wipe clear the nasal passages so the baby can breathe unobstructed. The mother should clean the birth fluids from her baby; in case she doesn't, use the clean towels from your birthing kit to get the kid or calf clean and dry and to stimulate blood circulation. Be sure the kid or calf gets on its feet and drinks colostrum, the first milk, providing immunity. Use the iodine to dip the navel cord.

Also be sure the cow passes the placenta within 24 hours and the doe passes the placenta in 12 hours. She should return to acting normal, rather than acting uncomfortable, soon after birthing.

We choose to calf-share, allowing our cows to raise their calves, for the ease it adds to our schedule.

Bottle Feeding and Dam Raising

With the point of home dairying being to have milk for your family, your cows' and goats' babies will need to share their mothers' milk with you. Have a plan for how to go about that before they're born. You basically have two choices: dam raising and bottle feeding. We have used both methods on our farm, and bottle feeding for the sake of milk production is a controversial topic.

Dam raising is less time intensive for you and more natural for the animals than bottle raising and makes for hardier kids with better mothering instincts. On the other hand, bottle raising leaves more milk for the kitchen.

Dam raising is, as it sounds, allowing the mother and baby to stay together so the baby can nurse. Keep kids and does together full time for the first week or so, and then separate them at night. Keep cows and calves together full time for the first month, then separate them for 10 hours at night, working up to a 14-hour separation. Milk the doe or cow first thing in the morning.

For bottle feeding, let the mother care for her baby after birth and let the baby get its colostrum, then remove the baby to a warm and dry area. Goats and cows are social creatures, so keep them in a group. Move them to outdoor pens when the weather is good and they're proving to be doing well.

Bottle babies need to be fed with their mother's milk, milk replacer, or a combination of them four to six times a day for the first few days, gradually decreasing to two or three times a day over the next week. We like to fill bottles with half milk and half milk replacer, when possible. Start feeding grain and hay when the babies are 2 weeks old—it may take them time to figure out they're supposed to eat these without a mature animal around to demonstrate. Solid foods stimulate their stomachs to fully develop.

You will likely have to bottle-raise a baby at some point, such as if a cow or doe dies or is otherwise unable to care for it. If a doe has more kids than she can raise, she'll choose the one or ones that she wants to take care of, which is why you need to be so attentive in the first day or so. We had a doe who died a few weeks after giving birth and another who had mastitis when her kids were a month old. We found it's much harder to train these older babies to bottles.

Dairy Goat Breeds

In looking at breeds, they all have their perks, and they all have their drawbacks. This list is a place to start, and you should research more extensively to choose those right for you. Spend some time with the breeds you think you'd like to see how well your personalities match. (Also read "Miss Congeniality" in this chapter.)

Alpine: This medium to large breed of dairy goat originated in the French Alps. Alpines are a high-production breed often used by commercial dairies for their high volume of low-butterfat milk. They're also known for being hardy in cooler climates.

Lamancha: The only goat breed originating in the United States, Lamanchas were recognized as a breed in the 1930s. They are easily known by their lack of external ears. They're calm, generally quiet, and good producers.

Their milk is so sweet and rich, I think it tastes like half-and-half.

Nigerian Dwarf: The smallest goat breed and the highest producer of butterfat, Nigerian Dwarfs are a popular choice for suburban homesteaders due to their short stature. Full-grown does are less than 24 inches (61 cm) tall and therefore easy to handle. Their milk is desirable for its extremely high butterfat, which lends well to soap making and cheese making. Of course, being so much smaller, they produce less than half of the volume of a full-sized breed, and as I learned when we brought them to our farm, milking their tiny teats is harder than milking "normal"-sized does. They're good for children who want to milk their own goats.

Saanens are the largest domestic goat breed, and they are often used in commercial dairy operations because of their high production.

Nubian: Nubians were our first goat breed, chosen for their beauty. Nubians are known by their pendulous ears and arched, Roman noses. They come in a variety of colors and can be solid, striped, or spotted (called "moon spots"). They are the beauty queens of the goat world and often come with the attitude to match. This breed originated in Africa and does well in hot climates. Nubians are notoriously vocal but are popular homestead breeds for their ample production of high-butterfat milk.

Oberhasli: This Swiss breed is known for being very amicable. Oberhaslis are docile and cooperative medium-build goats. They are a reddish brown with black markings and are notably quiet.

Saanen: The most popular breed kept by commercial dairies for their high production, Saanens are typically laid-back and cold hardy. These large goats produce a lower butterfat than the other breeds I mention here. You'll find them solid white or light cream with some black spots on their skin showing through.

Toggenburg: Known as the oldest goat breed, Toggenburgs originated in Switzerland. They are usually a soft brown with white stripes on their faces that have come to be called "Toggenburg markings," even when the markings appear on another breed. This very hardy breed thrives in cold climates. They are high producers of low-butterfat milk with amicable personalities.

OPPOSITE: *Lamanchas are known for their high butterfat milk and their generally quiet, docile personalities.*

TOP RIGHT: *Nubians are known for their pendulous ears and arched Roman noses.*

BOTTOM RIGHT: *Lamanchas do not have external ears, giving them a unique appearance.*

Production of Popular Dairy Goat Breeds

Breed	Daily Average Production	Average Butterfat %
Alpine	1.5–2 gallons (5.5–7.5 L)	3.5%
Lamancha	1–1.5 gallons (4–5.5 L)	4.5%
Nigerian Dwarf	1–2 quarts (1–2 L)	7–8%
Nubian	1–1.5 gallons (4–5.5 L)	5%
Oberhasli	2–3 quarts (2–3 L)	3–5%
Saanen	2–3 gallons (7.5–11.5 L)	3%
Toggenburg	1.5–2 gallons (5.5–7.5 L)	3.5%

Nigerian Dwarf goats are often a favorite of homesteaders living in suburban areas, due to their small, manageable size.

Hope is a four-year-old Jersey cow I purchased from my friend, Hannah. She was born on Hannah's farm, and I was grateful that she came with a built-in mentor. Consider this when looking for a cow to purchase!

Dairy Cow Breeds

Hope, a Jersey cow, and her 7-month-old heifer calf, Honey, were the first dairy cows on our farm. Honey is half-Jersey and half-Red Poll. When the sun hits her, she is the color of a jar of raw honey, a deep amber. Helen, another Jersey, came along not long after. She gave birth to Hallelujah, and then we had a whole herd of dairy cows and heifers on our hands.

As a homesteader, you are not limited to the top-producing dairy cattle breeds in the same way a commercial dairy would be. Production is important because you have to feed and care for these animals whether or not they're milking, but your family's needs may be met with a cow that produces less milk and has other qualities you want.

There are many interesting dairy cow breeds for the homesteader. These are a few of my favorites.

American Milking Devon: The American Milking Devon is a bright red, medium-sized, triple-purpose breed used for milk, meat, and draft. It's been developed to thrive on forage and in severe climates. We are bringing American Milking Devons to our farm, rotationally grazing them in cooperation with our neighbors. I look forward to using them as our beef option, while the bull can also be used to breed our Jersey dairy cows.

Brown Swiss: Originally developed as a dual-purpose breed for meat and dairy, Brown Swiss are very large, stocky animals now primarily raised for dairy. They have a longer gestation period than other breeds. Their color ranges from almost black to a light grayish-brown. Their deep-brown eyes and black noses make them endearing, and a dark tail switch and dark hooves are unmistakable Brown Swiss accents.

Guernsey: Guernsey milk is a golden color because it's high in protein and fat, as well as high in beta-carotene, the same pigment that gives carrots their orange color. The human body converts beta-carotene to vitamin A, which is good for your eyes, immune system, and skin. Guernsey cows are light reddish-brown to red and white. Their hooves, udders, tails, and muzzles do not have a pigment, so they're always white or cream colored. These ladies have a nice disposition.

Holstein: The largest of the dairy breeds, a mature Holstein cow weighs 1,500 pounds (680 kg)—nearly twice that of a Jersey—and stands 58 inches (147 cm) at the shoulder. These black and white cows are the most common dairy cows in the United States and certainly the most common in commercial production, with average production of 25,000 pounds (11,340 kg) of milk per year per cow. Holsteins can also be red and white or mostly solid black, white, or red.

Jersey: These beauty queens vary in color from a light gray or mouse color to a dark fawn or almost black. They are the smallest dairy breed and produce milk with the highest fat and protein content. Jerseys are the most efficient, making more milk per pound of body weight than any other breed. Jerseys put so much into their milk, in fact, that you can feed them more if they look like they need some weight gain, but a lot of the time, that just equates to more milk and not more weight gain. They are also known for high fertility rates and ease with calving, and they're better suited to hot climates than other breeds.

While dairying may be the most committed undertaking of the homestead, I assure you that you can do this. There will be times in your dairying experience that you'll wonder what you got yourself into, and there will be times that you wondered how you lived your life without it. If this is a dream that you're dreaming, prepare for it and pursue it.

Bringing home a dairy animal will change your life, both in the way you go about your day and in the way you look at the food you consume. Coaxing fresh, warm milk from your own cow or goat; pouring it into jars in your own kitchen; and taking it from the fridge hours later, cold and full of nutrients—it goes without saying that you won't look at milk the same way again.

Jersey cows are often kept by homesteaders because of their high butterfat milk.

Honey is a multipurpose hybrid breed calf.

THE FIRST-TIME HOMESTEADER

Hope has the minor defect of extra teats on the back of her udder. These can be clipped off at birth but are usually just a cosmetic issue. They do not produce milk.

Production of Popular Dairy Cow Breeds

Breed	Daily Average Production	Average Butterfat %
American Milking Devon	5 gallons (19 L)	4.0%
Brown Swiss	8.5 gallons (32 L)	4.0%
Guernsey	6.5 gallons (24.5 L)	4.7%
Holstein	9.5 gallons (36 L)	3.65%
Jersey	7 gallons (26.5 L)	4.8%

TOP: Excess milk can be used to fatten piglets, fertilize gardens, or feed chickens.

ABOVE: Choose your goat breed based on what you intend to do with the milk, not just on their appearance, no matter how cute you think their ears are!

Closing the Loop

On a sustainable homestead, "waste" from one area is beneficial in another. Milk is no exception. At times, your animals will produce more milk than you need in the home. Sometimes, you'll end up with milk that's not fit for humans, such as when a goat puts her foot in the milk bucket.

This milk is valuable in feeding other animals. You'll have the inevitable bottle baby who can make quick use of excess milk and save you from having to use milk replacer. Pigs and chickens love milk, too, offered in limited quantities to not upset their digestion.

In the garden, milk is a powerful 46-26-17 fertilizer. It can get stinky on a hot day, and its breakdown can tie up oxygen in the soil—so use it sparingly. Milk's microbial properties may also reduce virus and fungus on garden plants. A 20 to 30 percent solution of milk in water, sprayed up to twice a week, can prevent and reduce powdery mildew and mosaic viruses.

Why Butterfat Matters

Butterfat content measures the amount of fat globules in milk. Butterfat actually comprises more than 400 fatty acids. Butterfat content affects the texture (creaminess) and flavor (sweetness) of the milk and changes how the milk performs in cheese making and soap making.

Cow milk separates as it sits, the butterfat in the form of cream rising to the top, even in the refrigerator. It's easy enough to scoop the cream from the top and just as easy to shake the jar to incorporate it.

Goat milk, unlike cow milk, is naturally homogenized, so the cream does not rise to the top but stays evenly distributed throughout the milk. You can manually separate it, using a cream separator, so you'll have cream for your coffee, sour cream, and butter making.

In goat breeds, you generally have to choose between butterfat and volume. The breeds that produce the most milk produce the least butterfat. Goat milk has more of a "skim" texture than cow milk overall.

Butterfat content is a highly heritable trait. Consider the butterfat percentages in the breed tables in this chapter but realize there is variation within each breed. Production records for registered dairy animals include their milk volume and butterfat percentage, so look for those numbers if you're buying and breeding registered animals.

Butterfat production varies by stage of lactation, too. The lowest butterfat is found in early lactation and increases in later lactation. It's also possible your animals will produce more butterfat in the shorter day length of winter than they will in the longer day length of summer.

Miss Congeniality

Goats and cows have personalities. It's as simple as that. You may read that Nubians are loud or that Oberhaslis are very docile. Whatever may be typical of a breed, don't overlook the uniqueness of each animal.

Even within the Jersey breed of cows, Hope and Helen have opposite manners. We joke that Hope is a diva and Helen is a lady. They are the same breed on opposite ends of the personality spectrum.

My Lamanchas are mostly gentle and quiet, but the rudest goat I've ever had was a Lamancha. I've had typical Nubians who are vocal and pushy, and I've also had Nubians who were quiet little darlings who followed me around like dogs. I've had Nigerian Dwarfs that never challenged a fence and others who wouldn't stop escaping the yard and walking on the roof of the car.

Handling and socialization lead to more friendly and tame animals, but sometimes they just have more mischievous, dominant, playful, or loving personalities. These traits may be endearing or maddening, but it just is what it is. Their personalities are part of the fun of keeping them on the homestead.

Goats' personalities (just like humans) differ greatly. Regular handling, ample treats, and socialization lead to friendlier goats.

Raw or Pasteurized?

Milk straight from your cow or goat is considered raw milk. The milk you purchase in a carton in the grocery store is pasteurized, heated to kill most bacteria. It is not legal in most states in the United States to purchase raw milk for human consumption, but it is legal in every state for you to consume raw milk from your own animal. How you choose to consume your milk is a personal choice. Be sure you understand the complexities of milk, raw and pasteurized, as you make the choice for your family.

We drink our milk raw. It is a living food, full of enzymes and bacteria. Pasteurization inactivates nearly all bacteria, good and bad, including the *Lactobacillus* bacteria, which help us digest lactose. Pasteurization also changes the protein molecules, which doesn't seem to make a difference in cheese making.

The consumption of raw milk highlights the need to keep your animals healthy and follow a strict cleanliness routine. Some cow and goat diseases, like brucellosis and tuberculosis, as well as bacteria, can be transmitted via contaminated raw milk.

If you choose to consume pasteurized milk for food safety or other reasons, you can easily pasteurize your milk at home. Heat it slowly over a double boiler, stirring constantly until the milk reaches 165°F (74°C). Hold it at that temperature for 15 seconds, then put the pan of hot milk in a container of ice water and stir until the milk is cold. Put it immediately into the refrigerator.

A2 Milk

In looking at dairy cows, you'll find talk about cows with A2 genes. A2 milk comes from cows with a natural genetic variation. The A1 and A2 beta-casein proteins differ by a single amino acid. All cows produce at least some A2 beta-casein, but some breeds, including Guernseys and Jerseys, have predominantly A2 in their milk. Holsteins produce milk with roughly equal amounts of A1 and A2 beta-casein.

Some people say A2 milk is more easily digestible, and some studies support this. All goats (and sheep) produce A2 milk. If A2 beta-casein proteins are something important to your home milk production, look into lines of cattle that carry A2 beta-casein genes.

OPPOSITE: Before we had cows, I would visit my friends the Rhodes, and drink big mugs of cold, raw, A2 cow's milk. During one of those visits, I decided for certain that cows were in our future. It's hard to beat fresh living food from right off the homestead.

CHAPTER

5

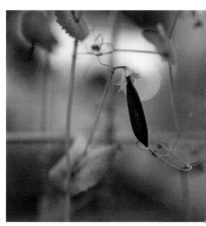

The Homestead Garden

When we started our homestead, I'll admit, the garden was an afterthought. I was busy dreaming of chickens scratching through the yard and a freezer full of meat we'd raised. I enjoyed the animal husbandry, but I didn't yet realize the integral role of the garden or how good it would feel to know I had all that food waiting for me outside.

When making your plans, don't put the garden on the back burner. Getting new animals is exciting, but the garden produces a lot of food for the money invested. The risk is low (seeds are easier to replace than goat kids, for instance), and the payoff can be high in food for your table and scraps for your animals' feed bucket.

I don't pretend to know everything about gardening. There are so many factors that go into growing. I am in USDA Zone 8a, so the varieties I plant and the time line I plant them on will be different than your varieties and time line. Know your USDA growing zone (or the equivalent for your country), mark your last frost and first frost dates on your calendar, and apply some critical thinking. Go in preparing to harvest wisdom first and foremost and garden success will follow.

"Cook things, eat them with other people. If you can tire your own bones while growing the beans, so much the better for you."

—Kristin Kimball

ABOVE: A garden that is out of sight is out of mind. Place it close enough to your home that you can easily water, harvest, and maintain it.

OPPOSITE: Create a garden you will enjoy being in and grow things you will enjoy eating. The pride of producing a space and food you enjoy is all the fuel you'll need to work through failures.

Garden Siting

Garden success begins with the right spot to put your plants. Find a place with at least 8 hours of direct sunlight. This exposure is considered "full sun," which is what most vegetables want.

Consider proximity to water access and how you will get water to your plants. Set yourself up to do well here. The task of dragging hoses to your garden beds becomes more of a burden as the season goes on.

The other side of water access is water drainage. Most plants do not want soggy roots. If you have a low-lying area where water pools, don't put your garden there, or find a way to divert the water around that spot before installing your plants.

Gardens are ideally flat, though in some areas, flat property is hard to come by. If a hillside garden is your only option, work with the contours of the land and terrace your beds to prevent erosion and protect your plants.

Take into consideration where your wind comes from as well. In flat and windy areas, consider windbreaks to protect your plants.

Think back to the permaculture zones I mentioned in chapter 1, "Making a Plan." Place your garden in a zone close to your center, as during the growing season, you'll work in it nearly every day. The easier your garden is to access, the more time you're likely to spend there and the better experience you'll have.

When you plan your food gardens—your perennial patch, in particular—have a general idea of your whole farm layout. We made the mistake on our first farm of planting our fruit trees in what ended up being a goat yard. Needless to say, those fruit trees didn't survive. At the same time, don't become so crippled by wanting to establish the perfect spot that you don't end up establishing it at all. You can often move your plants when needed.

Peach blossoms in early spring.

We live in USDA Zone 8a, which enables us to grow citrus like these kumquats.

The Perennial Garden

I love all the garden spaces on our farm, but the perennial food garden is the gift that keeps giving. A perennial food garden is a place of beauty. The term *perennial* means "through the years." Perennials are plants that you plant once, and they continue to return, bringing a harvest of food year after year. You get to watch a perennial garden grow and develop over time. What starts as a sparse-looking patch of tiny plants here and there will surprise you in its fullness the following season.

Anytime I'm asked the first step of establishing a new homestead, the answer is the same: Get your perennial garden started first. On our new farm, before we'd broken ground for the foundation of the house, before we'd raised a barn, and before we'd sunk a single fencepost, I began planning for the edible perennials.

It's important to know your growing season and conditions so you can determine which plants can overwinter as a perennial. For example, the bay laurel is a perennial in Florida but acts like an annual in Minnesota, where it will not survive the winter.

Many perennials need a period of growth and establishment before a first harvest is possible. Fruit trees, for instance, require 3 to 5 years of growth before they bear much fruit. A blackberry patch may produce a small amount the first year but won't show a significant harvest until the second. An asparagus patch requires 3 years before it can be heavily harvested. This is why I suggest getting perennials planted as a top priority. The sooner you plant, the sooner you eat.

Kitchen gardens and perennial food gardens are not the same, and you'll be happier keeping your perennials and annuals separate. Perennial gardens tend to be less maintenance than your annuals, as they generally don't require water after getting established and they often grow to a size that shades out most weeds. Kitchen gardens need to be cleared, replanted, and amended yearly. You may want to rotate crops and allow your animals in at the end of the season to clean up. Having to work around perennials is just too much hassle. This is not to say you can never plant annuals in your perennial patch, particularly in the early years when your perennial patch is still sparse and you have spaces to fill in, but plan on a separate space for the bulk of your annuals gardening.

Bare root plants and cuttings soak in a bucket of water with added rooting agent.

My Must-Grow Perennials List

Perennial plants include a variety of vegetables, fruits, and herbs. Start your must-grow perennials list, or use my list, and determine what you need to cultivate each plant. Most perennials start with cuttings, crowns, or rooted plants. Some want to be planted in the fall and some in the spring. Each has a different requirement for spacing and soil composition.

Asparagus: Asparagus is one of the delights of spring, ready to be harvested before you're even able to plant your annual plants. An asparagus plant can last 15 years, so choose your asparagus patch location wisely. Plant crowns (root systems) in the early spring. Don't harvest that first year; rather, let the feathery fronds grow and begin harvesting the next year.

Rhubarb: This garden beauty has huge, lush leaves on pink or red stalks. All parts of the plant except the stalks are toxic. Plant rhubarb crowns in early spring. They'll start producing significant stalks in the second year and should be productive for 5 years or more.

Raspberries: Plant your berries of all kinds in full sun for the best fruit production. Start with bare root or potted canes in early spring. They need annual pruning after fruiting. Unless you'd like to share your berries with the birds, you'll want to put up bird netting, and they're easiest to harvest when trellised. It'll take a year for raspberries to begin fruiting.

Blackberries: We had dark, delicious blackberries growing along the roadside near our first farm, and foraging them was a favorite activity. You can grow blackberries in a thorned or thornless variety on the farm, both from canes planted in the early spring. Like raspberries, they need pruning, bird netting, and trellising. Unlike raspberries, they may offer some fruit in their first season.

Blueberries: Of the berries, blueberries have the most peculiar growth preferences. Blueberries like acidic soil, whereas most plants do not. There are many varieties of blueberries suited for various climates with different fruiting habits. You need two or more varieties for fruiting. Plant young bushes in the spring. They will start their best production in their fifth year and can remain productive for dozens of years when well cared for. They don't require trellising but do need bird netting and annual pruning.

Mulberries: Mulberry trees grow quickly. Red mulberry trees easily reach 40 feet (12 m) tall, whereas black mulberries are smaller and may even grow like shrubs. As mulberries ripen, you're in a race against the birds and wildlife. These berries are delicate and don't keep long after harvest. Mulberry trees make great additions to a food forest, situated in full sun. They don't mind rough soils and require little water after getting established.

Currants: Once outlawed because they were thought to spread a tree disease, currants are delightful little berries that grow in clusters on thornless canes. It may take a bush up to 3 years to fruit, and one bush will yield about 1 gallon (2 kg) of fruit per year. Some varieties will fruit without companions, while others need an additional variety planted alongside. Plant bare-root and potted currant shrubs in the early spring, even when lows are dipping below freezing.

Jerusalem artichokes (sunchokes): These are prolific tuber producers from the sunflower family. Jerusalem artichokes are kind of like a cross between a potato and a water chestnut. Plant these tubers in early spring in an area that you don't mind giving up to a tall, spreading plant. I didn't harvest mine in the first season, as I wanted to give them a chance to get established, though you could take a conservative harvest after frost in the first year.

Walking onions: You know how I love a beautiful garden addition. In place of flowers, walking onions produce a cluster of tiny bulbs at the top of a stalk. The stalk falls over, and the bulbs replant themselves. It's like they "walk" across the garden. They're excellent green onions, and the bulbs growing in the soil don't get very big but can be used as regular onions. Plant these from bulbs in late winter or early spring.

Perennial Delights

A favorite feature of my first perennial garden was the herb spiral. It's a lovely garden feature and a useful one at that. Oregano, marjoram, thyme, and sage wove a fragrant culinary tapestry around the small stone structure that Miah built. These plants add pleasure to the garden and the kitchen.

Other perennial herbs I always have around are rosemary, lemon balm, and mint. (Be warned that mint is a bully and will take over the space. This includes other plants in the mint family, like lemon balm, stinging nettle, and catnip.) There are many varieties of these herbs, and their beautiful blooms and fragrances attract beneficial insects.

A perennial garden is a great place to tuck in ornamental plants as well. I have hydrangeas, hollyhocks, roses, peonies, dahlias, elephant ears, irises, and gladiolas adding interest and accents. My zinnias and cosmos, both annual flowers, reseed themselves throughout the space.

The Annuals Garden

Your annuals are your major food crops. Annual plants produce for one growing season and then die when the conditions become too harsh or they've cycled through their productive life.

Whether you are raised-bed gardening, no-till gardening, container gardening, or some other kind of gardening, every growing season begins with a plan. Without a plan, you'll find yourself starting late, rushing, and enjoying the process less. Your plan should be flexible because homesteading requires flexibility, but know what kinds of food you want to grow in your garden, learn when and how to start those seeds, and have an idea of what you hope to do with the harvest. You'll kind of figure out the rest as you go.

The gardener's quandary is having more things to plant than space to plant them in. Even after years of increasing my garden space, I still don't get to plant everything I want to plant. When making these hard vegetable and variety decisions, go back to your homesteading goals. Before you even open a seed catalog, ask yourself:

• How many people am I feeding?

• Am I growing food to supplement my family's vegetable intake during the main growing season, or am I planning to put by food to eat year-round?

• How much space do I have, and how much space does each plant require?

• What do we love eating? Especially when you're first getting started, stick with your favorites.

TOP: Savoy cabbage and its signature crinkled leaves.
ABOVE: A purple pole bean hangs from a trellis.

Choosing Varieties

Within each category of vegetable, fruit, or herb are varieties of each. Varieties differ in their ideal growing climate, their disease and pest resistance, their flavor, their appearance, and more. Talk to gardeners in your area about their favorite varieties to grow. As a beginning homesteader, set yourself up for the most success by growing things that are tried and true for your area and branch out from there when you've gained confidence.

Knowing I wanted to garden for beauty and flavor, I quickly delved into growing heirloom varieties. Heirlooms are older varieties, sometimes grown by homesteaders and indigenous communities 100 or more years ago. The stories behind heirloom seeds made a world of difference to me, and heirlooms' unique colors and tastes allowed me to discover how beautiful different plants can be.

For home vegetable gardening, you don't need to worry about whether you might be getting genetically modified (GM) seeds. You might find hybrids, open-pollinated seeds, and heirlooms, but GM seeds are highly regulated and only available for a small number of crops from a small number of seed companies.

Sugar magnolia tendril peas are my favorite snap pea variety.

How Many Plants?

Deciding how much to plant may be even more difficult than deciding what to plant. To start, understand how each vegetable grows and produces. If you're planting carrots, each seed will give you one carrot (if the seed even germinates). If you're planting peppers, each seed will produce a plant that grows a dozen or dozens of peppers. You need to plant more carrot seeds than you do pepper seeds to feed the same amount of people.

This is easier said than done, but I want you to start small. Maybe you start with a little salsa garden in a few raised beds so you can eat some of your food fresh from your backyard for a few months of the year. As you become more comfortable gardening and are willing to take more risks, you can expand to eat more of your diet from your garden for a longer part of the season.

It's not a far leap to start canning, freezing, and dehydrating your harvest, and then you're looking at needing a larger garden space. We grew about 80 percent of the vegetables for our family of eight in our 10,000-square-foot (929 m²) garden, but it took time to ramp up to that point. The work involved in gardening is not just the planting and cultivating; it's also the harvest and dealing

ABOVE: A garden grown for preserving will be significantly larger than one grown simply for fresh eating.

OPPOSITE: The tomato variety "Spoon" produces tiny, flavorful fruits.

with the harvest. You can't prepare yourself for putting by your first big harvest until you're in the thick of it, but you can temper your garden plans to reduce your initial overwhelm.

It's hard to say exactly how many plants to put in per person. One year I planted loads of cucumbers, and they produced so well that we canned enough to last for several years. We ended up feeding cucumbers to the pigs by the bucketful that year. The next year, I planted far fewer, and we had a terrible cucumber year. I actually had to buy cucumbers that summer. There are so many variables that affect how plants will produce. Planning for the right harvest is largely trial and error and luck.

Succession Planting

If you hope to eat from your garden beyond the typical summer flush, continue sowing seeds throughout the summer and into fall. This is called succession planting. A calendar, a notebook, or a spreadsheet on your computer will help you get organized for this one.

I always grow a few plants of Mexican Sour Gherkin cucumbers. They are hard to preserve but we love their novelty and bright flavor for snacking in the garden.

Starting seeds in waves for succession planting, rather than all at once, allows for the harvest to be staggered in ripening.

Rainbow chard grows well in a variety of climates and conditions.

Some crops are long-season crops. These include winter squash, sweet potatoes, and dry beans. You plant those once and harvest as they're ready. Other crops grow faster and have a shorter productive life. These are the crops you need to plant every week or few weeks throughout the season. Some of the vegetables and herbs that I succession plant include lettuce, green beans, summer squash, cucumbers, cilantro, dill, and basil. You may talk to small-scale farmers who put in multiple plantings of tomatoes and eggplant as well.

Year-Round Growing

Even in moderately cold winter climates, it's possible to grow a surprising amount of food in the fall and winter. Vegetables in the Brassica family (kale, cabbage, collards, etc.) and Chenopodium family (beets, Swiss chard), lettuces, carrots, cilantro, and root vegetables aren't bothered by the cold; even calendula and sweet alyssum are somewhat frost-hardy flowers.

Planting when the summer heat is at its peak, as you do for a fall garden, means paying closer attention to the water and shade that your seeds and seedlings receive. Some seeds need warmer weather to germinate, and some prefer it cooler. All of them need adequate and consistent water.

Growing into the seasons of frosts and freezes, an investment in row cover (also called frost fabric) can extend your harvest even further. You cover your plants with these long, narrow strips of fabric and weight down the edges of the fabric so it doesn't blow away. This lightweight material gives another few degrees of warmth to the plants while still letting light through and allowing the plants to breathe.

Potager Garden

The word *potager* comes from the French *jardin potager*, which translates to "kitchen garden." A small garden of this style provides vegetables, herbs, and cut flowers for the home. Potager gardens emphasize beauty while still serving the purpose of feeding your family.

When I started to design a garden I would love, I knew I didn't want to just put boring, straight rows in the ground. We started with raised beds and then built arched trellises. The gardens of my dreams mix beauty with practicality, and since the Middle Ages, monasteries have been doing just that with potager gardens. The monks and nuns, living in largely self-sufficient communities, looked to their gardens for food production as well as a place of respite and healing.

If you follow permaculture growing methods, you're familiar with diversity as a core principle of this holistic system. The monoculture crop fields so familiar to us as "agriculture" today are not part of a natural system. Nature and ecosystems are diverse and functional. Plants grown in community can provide nutrients, windbreak, shade, and support for their neighbors. The benefits of a diverse potager garden are numerous.

A beautiful, fruitful garden is good for the pantry and, more importantly, good for the soul.

Planting flowers may seem frivolous, but it's crucially important to support pollinators with diversity. The bonus is cut flowers for your home!

Planting flowers is an easy way to add beauty and creativity. Beautiful plants attract pollinators and other beneficial insects. That's the wonderful thing about this type of gardening: Not only is it good for the gardener, but it's also good for the garden ecosystem.

While the design of the potager garden matters less than the contents, these spaces tend to be less about straight, efficient rows and more about whimsy. Here, you can play with geometric-shaped beds and creative features more easily than you can in a large production garden. The symmetry of triangular tepees, even made of different materials, creates rhythm throughout the space. You can add height with trellises and tall plants, like sunflowers.

Every bed in my potager garden has some element of creativity, and I didn't spend a ton of money on them. Many of the artful touches in my garden are things that I found on clearance and at garage sales, plus items we had hanging around. This space feels personal to me, and I encourage you to make your space your own as well.

While you're adding your special touch to the garden, go ahead and put a chair out there. So often I go to my garden to sit, rest, and observe, and the next thing I know, I'm tying up tomato plants, singing, and enjoying myself. It doesn't feel like work, because it's my passion and this is where I come to relax. Create a space that you enjoy being in and you will find more success in growing your food.

Happy birthday, little radish seedlings.

Building Your Soil

Everything on the farm goes back to the soil. When I started gardening, the concept of soil health seemed too scientific and complicated. I promise you, it's not. Understanding soil health is the very basis of good gardening. In poor soil, your plants will grow slowly, grow spindly, and look sickly. You'll find higher pest pressure, too, because your plants will be less healthy. Healthy soil always grows healthy plants.

Think of your soil as a neighborhood. You don't want it underwater, and you don't want it bone dry. You don't want it to be compacted, with its airways and waterways clogged. You want a diverse population that works together.

There are four major types of soil: sandy, clay, silt, and loam. Each has its pros and cons in how well water moves through it, its ability to hold temperature and

nutrients, its resistance to compaction, and more. You should know what type of soil you have on your farm, but don't despair if someone tells you it's not "good" growing soil.

Soil Tests

A soil test is a simple way to learn what nutrients your soil lacks, those it has in abundance, the soil pH, and whether there are any heavy metals or other contaminants. Your cooperative extension office may offer soil testing. Test kits are also available from garden centers and big-box stores.

Soil tests are a helpful marker for soil health and will save you money in amendments. There's no reason to apply a high-nitrogen fertilizer if what your soil needs is potassium. Each plant has different nutrient and pH requirements as well, and a soil test can help you be more successful in growing what will naturally thrive.

If you want a more basic means of keeping your soil and your plants healthy, amend your soil with rich, organic matter, like compost or aged manure. Just be sure not to add any amendments with high amounts of one nutrient or another without first knowing your basic soil profile.

Soil Life

More important than the soil type is the soil life. The soil microbiome is teeming with life: billions of bacteria, fungi, nematodes, and more. These life-forms break down nutrients and minerals in the soil and make them available to plant roots, they degrade pollutants, and they help soil particles bind together to create soil structure that supports plant life.

Any soil can be amended to grow well. You just add biology (meaning life) to the geology (meaning the layers of earth). You might have soil that doesn't thrive immediately, but by adding compost and lawn clippings and using natural mulch materials, you are bringing life to and unlocking nutrients from the soil. I don't mean for this to be an oversimplification of a complex topic, but ultimately, that's the bottom line.

Various plants will help improve your soil as well. Simply by putting plant roots in the soil, you're reviving the life there. Deep tap-rooted plants, such as chicory and burdock, will descend several feet and bring up minerals and nutrients that are buried down deep. Legumes, like beans and peas, have nodules on their roots that work with bacteria in the soil to fix nitrogen for the next crop. All plant roots help break up the soil and create space for water and air to flow.

Your own work in the garden impacts the soil structure and life as well. Maintain the structure of your soil by avoiding walking on and compacting your growing areas, and don't work with your soil when it's too wet.

TOP: *This ginger harvest came from a handful of ginger broken into 1-inch (2.5 cm) pieces and planted in the early spring.*

CENTER: *Moon and Stars watermelon with its distinct variegation.*

BOTTOM: *Seed potatoes wait to be tucked into soil.*

A handful of soil contains billions of living microbes.

Compost

Compost is one of the most cost-effective ways to get biology back in your soil. In my original garden, I filled my raised beds with inexpensive topsoil topped with a few inches of good, rich compost. The first year, the garden did okay. After amending the soil like that for a few years, I could no longer tell where the compost ended and that original, cruddy soil started. The life from the compost spread through the soil, and we had a good earthworm population and a healthy microbiome throughout.

Ready-made compost is a convenient thing to purchase, but you'll pay for that convenience. Making your own compost is more cost effective and just slightly more work. Different types of manure, kitchen scraps, wasted hay, clippings from trees, leftover mulch, and garden waste break down into compost, their nutrients now available to feed your soil.

There are different ways to compost. The more effort you put into your compost pile, the more you'll get out of it. At its simplest, you can create compost by piling up organic matter and leaving it to rot on its own. Microbes are everywhere, they're hungry, and they'll happily break down this material for you. This is called cold composting.

Compost is most efficiently made when you balance carbon (also called brown materials) to nitrogen (also called green materials) at a ratio of two to one. This means for every 2 pounds (907 g) of wood chips, cardboard, and leaves (carbon), you need to add 1 pound (453.5 g) of food scraps, manure, and lawn clippings (nitrogen). You can spend a lot of time overthinking this ratio. What's most important is to provide a mixture of materials for the pile.

An efficient compost pile also needs water, whether from rain or your hose, and oxygen, which comes from regularly turning the compost pile. This carbon-nitrogen balance, water, and oxygen combo creates the right conditions for microbes in the pile to thrive and work quickly to create your garden gold.

Even in a cold composting setup, the action of breaking down organic materials creates heat. It's possible your compost pile can get so hot that it catches on fire. Keep this in mind as you situate your pile near any structures. A compost thermometer allows you to keep an eye on the temperature.

The Berkeley compost method is a "hot" composting method. This requires a large compost pile—at least 3 feet by 3 feet by 3 feet (1 by 1 by 1 m)—to maintain its ideal conditions. The compost pile needs the moisture consistency of a damp sponge plus daily turning to spread out the microbes and offer them oxygen. This pile operates at 135°F to 165°F (57°C to 74°C), ideal for the most active microbes. The 2:1 carbon-to-nitrogen ratio is important here to set up the hot conditions. The pile will be most active if you can provide it with materials in small pieces (for example, chunks of watermelon rather than an intact watermelon).

Things you should not put in your compost pile include meat, bones, fish, dairy products, grease or oil, seeds, and diseased plant material. It is unlikely your compost pile will become hot enough to safely decompose these materials, and you don't want them in your garden beds. Also, your dogs and local wildlife will make a mess of your pile if meat, bones, fish, and dairy products are on offer.

Rich life-filled soil is a treasure to the gardener.

Mulching

Looking around at spaces left to nature, you'll never see a patch of bare soil. In natural systems, soil stays covered, whether by vegetation or organic matter, like fallen leaves. All of those living organisms in your soil need cover for their ideal conditions. Mulching smothers weeds, holds in soil moisture, keeps the soil from splashing onto the plants, and keeps soils cooler in summer and warmer in winter.

I use straw mulch. It's easy to source and easy to spread. It can be composted at the end of the season or sometimes reused the next season. The downside is that sometimes it comes with viable seeds that germinate in your garden beds.

Grass clippings, shredded leaves, and wood chips are other good natural mulch options. Fabric mulch, with holes cut in it for your plants, are also popular. These are more expensive than natural options but can last a few seasons.

Feeding your family from your land is an effort, and a worthwhile one at that. You get the most bang for your buck in gardening, growing your own food for very little money. Fresh salads in the spring, bountiful zucchini dishes in the summer, sweet potato and pumpkin feasts in the fall, and jars of canned tomatoes for your comfort food all winter long—these are the promises your garden work holds.

Your homestead garden becomes your seasonal companion. When the excitement and hustle of springtime wear off, the busyness of summer's harvest and cultivation kicks in. As summer carries on and preservation work begins, the signs of fall start to creep in, giving signal that the main season is coming to a close. Fall's tease of just a little more nice weather carries cool-season crops a bit further until finally it's time for you and the garden to get a little rest.

When searching for straw mulch, look for a grower that does not spray herbicides.

Potting up seedlings in the greenhouse for the spring garden

Sunflowers draw pollinators to the garden and the whole head of seeds can be fed to chickens as a treat.

Pigs can prepare garden spaces by rooting through the soil, digging out rocks, and fertilizing as they go.

Homestead Symbiosis

Gardens and homesteads go hand in hand. If you're raising animals—whether for milk, meat, or just the fun of it—growing a garden comes naturally. Together, they help feed your family and form a closed-loop system where manures feed the garden and garden discards help feed the animals. The homestead garden grows in relationship with the rest of the farm.

When your harvest is finished, your poultry and livestock can enjoy what's left for them. You may tear out the spent plants and offer them with your animals' feed, or you may let your animals into the garden space at the end of the season for cleanup duty. Poultry can be useful in eating insects and slugs from your garden area, too.

The nutrients your livestock and poultry leave behind are perhaps the most obvious symbiotic element of keeping animals while gardening. Whether they're directly fertilizing the area you plan to grow in next or you collect their manure and bedding for compost, your animals nearly pay for themselves with this exchange.

If letting your animals into the garden, follow the basic food-safety guideline of not harvesting anything that touches the ground in that area for at least 120 days and not harvesting anything that grows without touching the ground for at least 90 days. Keep careful watch over your animals in this precious space, as you don't want them to compact or tear up your growing area—just remove the garden debris.

Before you break ground on your growing area, you could let your animals roam the space to graze and pre-fertilize the soil. Many farmers move their mobile chicken tractors over next season's garden area. The same food-safety rule applies.

While you're growing your family's food, your garden can produce livestock feed and forage as well. You won't be able to grow your own hay for cattle without dozens of acres of hay fields, but you could grow a row of forage to cut fresh for your rabbits, corn to crack and supplement as scratch for your chickens, and more.

I directly sow rutabaga seeds into the garden six to eight weeks before my first fall frost date.

Seed-Starting Advice

Sometimes it's best to start seeds indoors, and other times, it's best to direct-sow them into your garden. One benefit to starting seeds indoors and putting them out as transplants is that by the time you put them outside, they're larger and can better fend off pests. A downside is that some types of plants, like winter squash and beans, don't like their roots disturbed and may initially grow more slowly after transplanting.

Seeds that produce just one vegetable each, such as carrots and corn, make more sense to sow directly into the garden, whereas those that produce whole plants full of fruits and herbs may be more worthwhile to start ahead of time. Some plants that take a long time to produce, like tomatoes and peppers, may need to be started inside for that extra growing time, particularly if your growing season is on the shorter side.

Starting tomato seeds indoors means I'll harvest fruit much earlier in the garden season.

Look to your local university- or governmental-based agricultural office and to various seed catalogs for sowing guides. These will help you determine the best time to start your seeds, whether you're doing so in seed cells indoors or direct-sowing outdoors.

Starting seeds indoors requires a loose potting mix and a container that allows water to drain through. Bagged compost and garden soil tend to be too dense for starting seeds. The light, fluffy texture of a good seed-starting medium allows the seed to sprout roots and the roots to easily penetrate the soil. The seed coating itself contains the nutrients the seed needs to burst forth, and after a few days, the roots will need to source nutrients from the soil. Many commercial seed-starting mixes do not have nutrients, so you may need to mix in worm castings or compost. Pay attention to the label on the bag.

Plant seeds twice as deep as the seed is wide. In the case of radish seeds, that's very small; you just dent the soil, place the seed, and cover it. With peas, these are larger, and you need to poke a hole to drop the seed into. This is true whether you're direct-sowing or starting inside.

Seeds need consistent watering to germinate, of course, but too much water is not good for them. Small seeds, like carrot seeds, can even wash away in a heavy rain. Pay attention to the forecast when direct-sowing. There's a difference between a light germinating rain and a devastating downpour.

Whatever method you use to start seeds, be sure to label them! Tiny collard greens and tiny cabbage plants look similar, and you simply won't be able to identify different types of tomatoes until they start fruiting.

Start seeds in a fluffy potting mix.

Label your plants so you don't forget what you've planted!

My friend, Wes, grows potatoes in upcycled food-grade barrels.

Saving Seeds

Part of self-sufficiency is producing your own food without outside inputs. Saving seeds from your plants each season has compound benefits. You don't have to spend money on new seeds each year, and the seeds you save from your own plants are more well adapted to your growing conditions.

Some plants are easier to save seed from than others. Tomatoes, peppers, beans, corn, sunflowers, winter squash and pumpkins, and cucumbers are excellent beginner seed-saver options. Their seeds are large and obvious. It's harder, but not impossible, to save seeds from plants that don't fruit, such as carrots and greens.

Many a seed saver's downfall is planting too many varieties of the same vegetable. Most garden plants cross-pollinate, but you may want to collect seeds that will breed true to their parent. (Tomatoes, flowers, and beans do not generally cross-pollinate.) Be careful to either plant only one variety of the seed you're saving (just one type of cucumber, for example) or plant different

varieties very far apart—100 yards (91.5 m) or more. I don't mind accidental crosses, so I never worry too much about this. The worst thing that can happen is a new hybrid!

Another option for keeping your plants from cross-pollinating is to cover the blossoms with mesh bags and hand-pollinate the flowers. You don't have to go through this effort for your whole garden—only for the plants you expect to collect seeds from. Just one plant can produce hundreds of seeds.

The seeds you save are genetic reproductions of the plants in your garden, so choose the best plants for your saving efforts. They ideally will be free from disease, hardy under pest pressure, and very productive.

OPPOSITE: In each fruit we harvest there are hundreds of seeds, which represent thousands of future fruits. The multiplication of the garden is truly astounding!

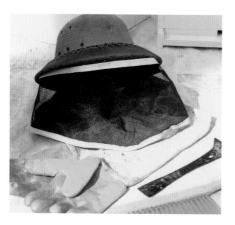

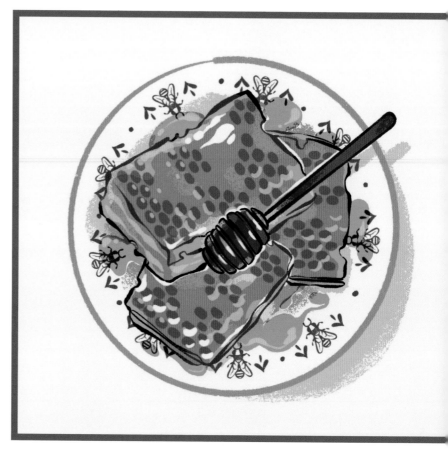

Keeping Bees

I count bees among the most valuable, not to mention interesting, livestock on a farm. While there are thousands of different bee species, honeybees (*Apis mellifera*) are a primary pollinator of our fruits and vegetables. The more bees visiting the garden, the more productive the garden. This is true for food crops the world over, and as bee populations dwindle, small-scale beekeeping becomes even more important. In addition to this invaluable pollination service, your bees offer sweet, rich honey and beeswax to use in candles and crafts.

Whereas your work with more traditional livestock requires an everyday rhythm, keeping bees takes more of an observational approach. You create a home the bees will love, protect them from predators and other disruptions, and ensure they have food sources—mostly provided by nature. Meanwhile, the bees clean up after themselves, take care of their own breeding, and provide you with benefits you can't find elsewhere. Plus, they're fun to watch. It's not a bad trade.

We started our beekeeping with three hives. As our hives were multiplying, I asked Miah if he was going to fall for bee math, the way it's easy to fall for chicken math. He assured me he was not, but in no time, we had eight hives—which is four more than Miah said he wanted. With 60,000 adult bees on average in each hive, our bee yard was housing nearly a half-million bees. These insects, it turns out, are also collectible.

When we moved our farm across the country, we gave our bees to friends, as we didn't want to stress them with that move. As I write this book, we are making plans to reenter the world of beekeeping—and I can't wait!

Beekeepers foster a great community of friends and mentors. Like other farm pursuits, you need some mentorship to get started. We continued to call on our beekeeping mentors throughout our time with the hives, and I know we will do so again as we get reestablished. Many areas have their own beekeeping groups, agricultural offices often have resources, and the folks at local beekeeping- and honey-supply stores are troves of information. Reach out and learn all that you can about this fascinating work.

"He is not worthy of the honey-comb, that shuns the hives because the bees have stings."

—William Shakespeare

Inside the Hive

The inner workings of a hive seem to be a mystery until you get in there and look around. It's a fascinating world. There are various methods of beekeeping, yet down deep, bees are still bees.

Types of Hives

Langstroth hives and top bar hives are the two main types of hive structures, though there are others. We keep bees using Langstroth hives, which are best for beginners. Top bar hives are the almost-V-shaped hives. They yield less honey, yet they're lighter to deal with.

Langstroth hives: Named after their creator, Lorenzo Langstroth, these hives consist of a series of stacked boxes covered by a lid. The number of boxes depends on the size of the bee colony. Boxes that house the queen and brood (larvae and eggs) are called brood boxes; boxes for honey production are called supers. The boxes come in deep, medium, and shallow sizes and hold 8 or 10 frames. Bigger is not necessarily better, as a full deep super can weigh as much as 90 pounds (41 kg), which is a lot to manage. The whole setup rests on a hive bottom board, which allows bees to come and go through a gap in the front.

The bees construct their wax comb within the frames inside the boxes. You can provide them with frames that have a wax comb already in place, making less work to get established. Shallow and medium supers have the same size frames, and deep supers have a larger frame.

A honeybee enjoys a borage flower in a potager garden.

These dimensions are universal, making it easy to purchase or build replacement frames.

The measurements of a Langstroth hive are based on "bee space," which is the space that is large enough for a bee to fit through but small enough that a bee will not build a comb within it. If this space is any smaller than "bee space," the bees seal it up with propolis (a resin-like material made by bees) for security and comfort.

TOP: *Place weights on top of hives to secure them.*

ABOVE LEFT: *Langstroth hive boxes come in deep, medium, and shallow sizes. Make sure you can easily lift and maneuver whichever size you choose.*

ABOVE RIGHT: *Hive boxes can be painted to suit the beekeeper's taste.*

TOP: A horizontal top bar hive

ABOVE: Top bar hives can vary in size and are less structured inside.

Top bar hives: This hive style is a single horizontal box with tapered sides and a roof with a series of bars hanging just underneath. The bees build their honeycomb from these bars, straight down. This is less structured than the frames of the Langstroth hive and easier to harvest, but they yield less honey. Top bar hives are popular because inspecting the hive frame by frame is less disruptive to the bees than having to move supers.

Unlike Langstroth hives, the dimensions of the top bar hive aren't universal. On the plus side, you can build your own without exact measurements. On the downside, parts you order from a beekeeping supply store aren't guaranteed to fit your hive.

The Secret Lives of Bees

Miah describes the hive like a kingdom. You have the queen, and everyone serves her and has a job. Bees are the most studied creatures on the planet, and it's no wonder. Their self-sufficient system puts tens of thousands of bees to work without a hitch. They provide for their own networks of communication, home construction, defense mechanisms, nutrition, and more.

Each colony has one queen (two queens when they're preparing to swarm or replace the queen). She is the only female responsible for egg laying. One queen can lay up to one million eggs in her 5-year lifetime.

Drones are the male bees. They mate with the queen and then die.

The worker bees are the females that make up the majority of the colony. They feed the queen, prepare the honeycomb cells, care for the eggs and larvae, guard the hive, and generally keep the hive clean and operational. Workers do the foraging and bring back nectar and pollen also. They can lay unfertilized eggs when the hive is under stressful conditions, but those eggs will hatch drones only, which can't sustain a colony. They live from 6 weeks to 6 months, depending on the time of year.

Part of the ingenuity of the bees' system is to keep separate the reproduction activity and the honey-making activity. This is especially obvious in the Langstroth hive because the bees use separate boxes for each activity, but you can see the separation in any hive.

Bees, heavy laden with pollen, return to the hive after foraging.

When beekeepers talk about the brood, they're talking about the reproductive activity cells: the eggs, larvae, and pupae in the honeycomb cells. This is the part of the hive that maintains its population, and you can learn most of what you need to know about your colony by examining the brood (more on that to follow).

Nectar, Pollen, and Propolis

There are jokes about honey being bee vomit, and while that's not quite accurate, it's close. When a worker bee goes foraging, she consumes nectar from flowers and regurgitates it to another worker bee waiting at the hive entrance. The second worker bee regurgitates this nectar into a honeycomb cell. With each regurgitation, the bees add enzymes for preservation. The worker bees in the hive dehydrate the nectar by fanning it with their wings until it becomes honey, they cap the cell with wax, and the honey is ready for long-term storage. Bees consume honey for carbohydrates and minerals, which is most important in daily flight and activity.

As bees search for nectar, they collect pollen. They go from blossom to blossom, pollinating the flowers by spreading the pollen, and carry the pollen back to the hive using pollen baskets on their legs. Bees consume bee bread (pollen mixed with bee enzymes) for protein, fats, and vitamins. Pollen is most important for bee reproduction and development.

Also in their pollen baskets, bees collect resin from plants and trees. In the hive, this resin is called propolis, used as a construction material of sorts. They patch crevices smaller than "bee space" and smooth rough surfaces. Propolis is the sticky substance they use to seal the lid on the hive and the reason you need a hive tool. Propolis has medicinal properties as well. Bees collect extra resin for propolis to fight infections in the colony, and humans use propolis in holistic medicine.

A Beekeeper's Gear

For as little as you need to work with your bees, there is an investment in equipment. There are the hive structures themselves plus protective gear, a smoker, and other hive tools.

Your Bee Suit

People always want to know if you get stung when you keep bees on the farm. Bees have stingers to protect themselves and their honey, and yes, of course you run the risk of being stung by a bee. Know that the bees don't want to sting you, though. In fact, they die when they sting you. This is their last-ditch defense mechanism. You'll learn your bees' behavior and will become comfortable working around them.

Proper gear helps protect you from the occasional bee mishap. (If you've followed us on YouTube, you know that Miah's eyelid had an unforgettable-looking bee sting, which happened after he took off his suit.) Some people work their hives without a protective suit or gloves. We are not those people. You should do as you are comfortable.

If you're thrift minded, like I am, you might struggle with spending a lot of money on your bee suit, but if you choose to wear protective gear, don't skimp. Find a suit that fits you well, allows proper movement and range of vision, and is comfortable, because sometimes you'll be wearing it for a few hours in the hot sun.

Protective gear could include gloves, a veil, a jacket, or a full suit. Rookie beekeeping mistake: Don't beekeep in yoga pants. I learned this lesson so you don't have to. Durable materials, like denim or canvas, are more protective.

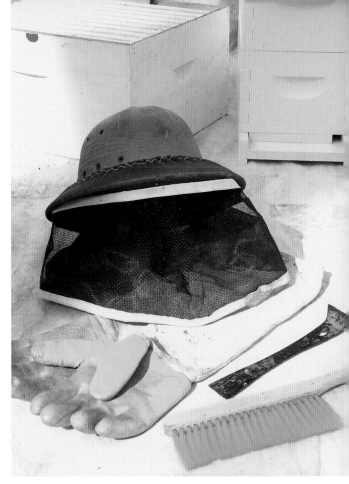

Proper protective gear and tools guarantee safety and ease for the beekeeper. Though some experienced keepers forgo using gear, I suggest anyone starting out be properly equipped.

Hive Smoker

A smoker is a metal chamber designed to direct cool smoke into a hive. The smoke makes the bees go further into the hive and chill out, making working in the hive easier. The smoke also masks pheromones, including the one that bees give off when they sting, which is an alarm signal for others to sting.

You'll see different versions of smokers. A good smoker will get rolling in 5 to 10 minutes using straw, wood chips, or other dry, flammable, organic materials and will stay hot for a long time. A warning: Smokers get very hot.

Hive Tool

You may guess from the name that a hive tool is a general tool for working in the hive. It's invaluable in allowing you to pry apart frames and remove the lid from a hive when the bees seal the cracks with propolis. You can use it to remove comb that bees have built outside of the frames, too.

Hive Bells and Whistles

This chapter is covering only beekeeping basics. You can dive as deep into this aspect of homesteading as you can imagine. You may find others recommend you have things like hive beetle traps, a queen excluder, entrance reducers, mite boards, and more. You'll learn about all of these tools as you go.

Sourcing Bees

You have a few choices for finding bees to bring to your homestead, both wild and cultivated. I recommend starting your beekeeping journey with a well-planned purchase of a bee nuc (see the next section). You'll have opportunities to collect swarms and to use simple bee packages as time goes on and you have more experience.

Have your hives set up and know how to use your smoker before bringing home your first bees. When they arrive, you'll open their box to inspect them. If they're mellow, you might not need the smoker, but keep it nearby because you want your first experience to be a good one.

Introduce your new bees when some forage is available so they can get started building their own self-sufficient colony right away; otherwise, you'll have to provide food sources for them. Springtime is the best time for this.

A Nuc

Packaged in a box, a nuc (nucleus) is an already-established colony that arrives ready to be installed into your hive. The adult bees, a queen, a few Langstroth frames, comb started with brood, nectar, and pollen are ready to go. Your nuc should come with installation instructions.

You might find nucs for sale from neighboring beekeepers or from a bee-supply store. Nucs can be sent through the mail by a shipper who knows what they're doing.

You have the option of purchasing three, four, or five frames with your nuc. Install these frames with the bees on them into the center of the box, and place the empty frames on either side, using smooth and slow movement.

In the nuc, honey frames and brood frames are in the same box, and you can start your hive off this way as well. Depending on how full the frames are when they arrive, expect to add a super to the hive and split the honey and brood frames into their separate spaces within a week.

A Package

If you already have your hives, equipment, and some experience, you may only need a bee package. You might be able to get a package from a local beekeeper who is splitting a colony, or you can order one from a beekeeping supplier and have it shipped to you. Either way, your package will include the queen bee and a population of workers and drones but no frames, honey, or brood.

When bees fan at the entrance of their hive with their tail-end facing away from the hive, they are releasing a pheromone called the Nasonov pheromone, which is used for guidance. This is used to signal for forager bees to return home.

A Swarm

The most exciting way to populate a hive is by capturing a swarm. Miah has done this a lot. Swarms are how we grew our bee yard so quickly. The chance of swarms is a good reason to keep an extra hive on hand.

The first time our bees swarmed, I was working in the garden on a spring day and heard this noise swell. At first, it sounded like a string trimmer—and it kept getting louder. I looked up and saw a massive cloud of bees hovering above the hive and just started yelling, "Miah! You're going to want to come here!" It's a surreal experience.

After a telephone call with our beekeeping mentor, Miah suited up, caught the swarm, and gave it a home in an empty hive we had on hand. We knew these bees had swarmed from one of our existing hives, so we knew what we were getting with this bunch. If you catch a swarm from unknown origins, you may end up with a hive of bees that have a bad temperament. It's the risk you take for free bees.

A honeybee swarm happens when a beehive gets overfull or the bees experience a major stress. The old queen gathers up half of the bees and leaves, usually gathering on a tree branch or the side of a building. While most of the bees from the swarm are clinging to one another in this group, scout bees go off looking for a more appropriate place to nest. It's during this time that you must work fast to gather the swarm and relocate them to your empty hive.

One of the swarms that Miah collected was on a tree branch about 6 feet (2 m) high. He shook the branch so the bees fell into a cardboard box, put the lid on the box, and dumped the box into the hive. You wouldn't expect

OPPOSITE: Here, a new nuc is establishing in an empty hive. ABOVE: Provide protection to your hives to deter predators.

it, but bees fall rather than fly in and out of boxes easily. We also use lemongrass oil (which bees are very attracted to) in the box to lure swarms.

You may be able to cut down the branch where the bees swarmed and shake them directly into a hive. Sometimes you can just leave a hive box next to the swarm and let the bees find it themselves. Every swarm experience is different, and each should be approached with thought and caution. If the bees are aggressive or the weather or location conditions are bad, let the bees go and wait for the next swarm.

You want to get the queen when you collect the swarm because the rest of the bees will follow the queen. If you get the queen into the hive on your first try, the remaining swarm stragglers will go to the hive to find her. (You'll never collect the full swarm in the box on the first try, but you should be able to get most of it.) If the queen is still out in the wild, the population you caught will leave the hive again to return to their queen, and you will have to start over.

You can also reduce the swarms from your own hives with regular hive inspections. When you see a new queen developing or notice the boxes getting full, split your hive into two colonies to prevent a swarm. Splitting a hive is less of an adrenaline rush but certainly easier than catching a swarm.

Siting a Bee Yard

Bees are a little bit picky in wanting just the right homesite. They enjoy their peace. Not only that, but you want to place your hives where they're convenient for you as well.

To start, have easy access to the hives, because the equipment is heavy and not fun to carry long distances or up hills. You won't work in the hives every day, so you can place them in a zone farther from your working center, reflecting on the permaculture zones from chapter 1, "Making a Plan."

It makes sense to put hives near the garden, as you want the bees to spend their time there. Bees can travel miles to reach nectar sources, though, so don't stress about having the hives and the garden right next to one another.

Bees need water to cool the hive. A bird bath or even a dog bowl in their general area will do.

Place the hives with the opening facing east, ideally with a natural or constructed buffer from the prevailing wind behind them. Varroa mites, or hive beetles, target hives in full shade, such as those in forested areas. Some shade is good, like trees for a break from the summer-afternoon sun, but sun exposure is more important.

Elevate hives to protect them from invading insects, wildlife, and rainwater. Cinder blocks will do, or you can build or purchase a hive stand.

Regular, loud activity around the hives will annoy the bees, which is not good for you or for them. Running a string trimmer or a lawnmower directly alongside the hives isn't a great practice. Roadways as well are an issue because of pollution, noise, and bees having to fly across traffic. (They don't know to look both ways.)

Livestock pastures are okay for beehives if you fence off the hives. Your bees will not bother your animals without provocation, but a curious animal can damage the hives and get stung in the process. Do not keep animals penned up near your hives, because if a disturbance were to cause your bees to go after an animal, that animal needs to be able to escape.

TOP: Bees need shallow water sources nearby.
ABOVE: Place beehives in an area without excessive activity.

We started out with beehives alongside our rural road, next to the greenhouse and adjacent to the garden. It was obvious from the beginning that the bees didn't think it an ideal spot. Our second bee yard was away from the road and had a privacy fence in an L shape to provide windbreak. We put down landscape fabric and gravel to cut down on maintenance around the hives. The bees were happier there. We placed the hives in a horseshoe pattern and gave them 5 feet (1.5 m) in between to prevent bees from drifting (returning to the wrong hive).

If you're keeping bees on a homestead in town, check your ordinances. Place your hive 10 feet (3 m) or more from your property line, and orient the hive opening away from your neighbors so they don't get in the way of the bees' flight path. Let your neighbors know about your homestead addition so they're not surprised by visitors and so they can rethink any chemical lawn care practices. Sweeten your relationship with an occasional gift of honey.

Natural vs. Conventional Management

As with all things on our farm, we prefer to manage our honeybees with natural methods. Sometimes we'll step in with conventional management in life-or-death situations where natural means aren't working.

Inspecting Your Hive

Plan to do hive inspections on sunny, calm days, because bees' attitudes match the weather. You want to check on brood, pollen, nectar, and honey in the boxes. You can observe the overall health of the hive with these check-ins, and you'll learn how to read your hives over time.

Be gentle with the frames and bees during inspection. You don't want to crush any of your bees, both because they're your bees and because you don't want to anger the hive. You especially want to use caution to not harm the queen.

Pull out each frame, one at a time. Especially on cool days, keep the frames out of the hive for just a few minutes at a time. Replace frames in the same order you found them.

Feeding bees with a sugar-water mixture in winter may be necessary for them to thrive.

Feeding through Winter

Bees put by all this honey to feed themselves through the winter, when nectar and pollen are hard to come by and it's too cold to fly. (Bees don't leave the hive when it's less than 57°F [14°C] outside.) We take a good deal of their honey for our own uses and allow them enough to get through a "normal" winter.

As our weather becomes less predictable and springlike periods pop up in the middle of winter, honeybees are just as confused as we are. They'll take flight in search of nectar sources and find none, returning to the hive worn out. It's during these times that we supplement bees with food. Sugar water and sugar bricks are easy do-it-yourself bee energy sources in lean times, and you can purchase these from beekeeping-supply stores as well.

Controlling Mites

Varroa mites are the bane of the beekeeper. These tiny insects reproduce in capped brood, and they transmit viruses and weaken the bees. It's essential to control mites before winter sets in, as the colony needs to be strong before facing harsh weather and fewer food sources.

Chemical and organic control options exist, and none is perfect. Trapping mites, using a screened bottom board that the mites will fall through, a mixture of essential oils, and sprinkling powdered sugar on the bees in the hive are also imperfect (but more natural) remedies.

You can eliminate mites by removing your queen, which eventually does away with the brood that the mites were thriving on, then returning a queen after the mite population has gone. Of course, starting with a bee colony that's shown its resistance to mites is helpful, too.

Protecting Your Hives

As wild as bees are, even they require some protection. Animals, chemicals, and sometimes even people can pose a threat to beehives.

We have lots of reasons for not wanting to spray chemicals on our farm, and the value of our pollinator friends is one of them. Bees are very sensitive creatures. Even chemicals that don't come in direct contact with the hive can affect the whole colony.

Before bringing home bees, alert all utility companies and your county that you are a beekeeper. We lost several hives when a utility company sprayed utility easements with glyphosate (the weed killer more commonly known as Roundup), causing our hives to abscond. We complained and made calls to educate them, but it didn't bring back our bees. Make sure you are on any no-spray list available, and if you see anyone spraying public areas near your hives, be bold and interrupt them, call their supervisors, and protect your bees.

If bears live in your area, they might go after your bees' honey. Just like cartoon bears, real-life bears have a sweet tooth, too. Ratchet straps can keep out bears but might not prevent them from knocking over the hives while trying to get in. Use an electric fence as an additional barrier.

For hives situated in an area in public view, curious people, as well as vandals, can become a problem. An electric fence can be useful here, as well as signs. You might have to go so far as to put up cameras—or fake cameras at least.

Beekeeping is a topic full of strong opinions. My advice is to find your way and make peace with it. You will always have someone agreeing or disagreeing with you.

Post signs and call utility companies and municipalities to request no toxic pesticides are sprayed near your home.

Bees, like all living things, have a drive to survive. The world's bee populations are in trouble, and without them, our food system cannot survive. Commercial pesticide use has hugely impacted bee colonies. We need bees, and therefore, we need beekeepers. It would be better to have many beekeepers raising bees imperfectly than a few who get everything right.

There is loss involved in the learning curve, and even experienced beekeepers lose hives. Just get started, and as you persist, you will be rewarded with wonderful honey and the great reassurance that you are contributing to the solution of a very real and present problem.

Honey will vary based on the time of year and the types of blossoms the bees forage.

A Bit about Honey

The color and taste of honey depends on the pollen source. Based on the plants blooming at the time your bees were collecting pollen, your honey may taste earthy, nutty, herbaceous, or fruity. It can range in color from deep amber to translucent gold. Buckwheat honey smells more earthy and tastes like molasses, yet clover honey is sweet and floral. This seasonal variation is a treat.

The honey that comes from your hive, like the milk that comes from your cows, is raw. You'll want to filter your honey to remove debris, but processing it is up to you. Raw honey contains natural enzymes and antioxidants. Pasteurized honey is heat-treated, which kills some enzymes but keeps the honey from crystalizing for a longer period.

About the honey that's been in your cupboard for months that's turning to crystals: It's perfectly good. Some honey crystalizes more quickly than others because nectars from different sources have different levels of sugars. Nectars with a lot of glucose are more likely to turn out honey that crystalizes. Let your honey jar sit in a bowl of hot water or, if it's in a glass jar, leave it on the stove as you're cooking and the crystals will melt back into liquid honey.

A Home for Native Bees

The honeybee is just one of thousands of bee species. While they're now vital pollinators for our food crops, they're not native to North America. Honeybees were brought here from Europe in the 1600s.

Bee houses and structures can house native bees, which are important pollinators for the garden.

You can welcome native bees to your homestead, too, because the more pollinators, the merrier. Native bees will find your farm welcoming with the same things your honeybees enjoy: a variety of plants blooming throughout the seasons, a water source, and shelter.

Unlike honeybees, a lot of native pollinators do not live in colonies. Their homes are smaller and less complicated, and because we're not managing them for honey, we can set them up and leave these bees to themselves.

Snags (dead wood) and hollow plant stalks are favorite native bee habitats. Leave some of your garden plants standing at the end of the year—perennials, especially—to give these insects a place to nest over the winter. Remove the debris in late spring, after insect eggs hatch.

Tree stumps and logs left on your property give wood-boring bees a place to live, too.

Many native bees live in the ground, pointing to another benefit of no-till gardening. Insects thrive especially in mulched areas and those along tree lines.

It's easy to construct native bee boxes as well. You've seen these: They look like bundles of straws stuffed horizontally into a box. That's kind of all they are: bundles of bamboo canes or other hollow plant materials in varying sizes for different native bee species. Secure these in a box and hang the box from a tree or post with full or partial sun exposure. Alternatively, if you have scraps of untreated wood (4x4s or larger), drill deep holes of different diameters into the blocks for the same effect.

Because native bees, not to mention other pollinators, live everywhere on your property, be thoughtful about any chemical use at all. Spraying one area, even if you're not growing food or keeping animals or bees there, affects the whole farm ecosystem.

Colony Collapse Disorder

In 2006, disturbing news about widespread honeybee die-offs began rolling in. While it's not as prevalent now, Colony Collapse Disorder (CCD) still happens and remains an unsolved mystery. It's important to know that your bee colonies may thrive and they may falter, but a hive failure doesn't immediately indicate CCD.

What CCD hives have in common are that most of the adult bees disappear; there are few, if any, dead bees in or around the hive; the queen, brood, honey, and pollen are still there; and Varroa mite levels aren't high. It's like the bees just disappeared. The prevailing theory for the alarming outbreak of CCD is pesticide use. Bees forage plants that have been sprayed with toxic chemicals and bring the chemicals back to the hive.

Any effort we can make to protect our pollinators is an important one. CCD is not yet predictable or solvable, so it's up to us to be good stewards of the honeybees in our care.

Aggressive Hives

That honeybees have stingers invokes such fear in people, but this fear is largely unfounded. A bee is not out looking for a target. (Remember, a bee dies after it lands a sting.) Mostly, honeybee colonies are docile. We did have one hive that was just mean. Miah's face sting and one incident I had with multiple stings on my leg came from working with that aggressive hive.

There is a difference between having honeybees that are aggressive by nature and having a hive that becomes aggressive in a disturbance. When a hive is disturbed, guard bees let off a pheromone that puts soldier bees on alert. Whether they come out and sting or just chase off the offender depends on the aggressiveness of the hive.

If you end up with an aggressive hive, removing their queen and adding a new queen can help. The queen is in control of the colony's genetics, so within a month of requeening, the collective attitude of the bees can turn.

It's important to deal with an aggressive hive quickly rather than to let the problem persist, especially if your bees are in the city or suburbs. An aggressive hive is a nuisance and can be a danger.

Highly aggressive hives are not the norm and can be requeened to achieve a more workable demeanor.

CHAPTER

7

Resourceful Living and Natural Remedies

Harvesting your own vegetables, communing with and receiving milk from your animals, and having meat at the ready in your freezer all bring with them an excitement and a feeling of accomplishment. Producing food as a homesteader is just one part of having the resourceful lifestyle that will allow you to leave the rat race.

A penny saved is a penny earned. This counts for shopping for clothes secondhand, investing in equipment that lasts, and making your own health remedies and cleaning supplies. The time and energy you invest in a resourceful homesteading life allows you to spend less time and money at the grocery store and less energy looking outside of your community to meet your needs.

Shifting from the typical rat race into a mind-set of having and using only what you need requires a long, sometimes slow, adjustment. Take a moment to assess where you're spending money unnecessarily. What things are you purchasing that you could make from scratch? What materials do you have on hand that you can reuse? What is complicating your life more than simplifying it? These questions are the beginning of a back-to-basics lifestyle.

"The secrets are in the plants, to elicit them you have to love them enough."

—*George Washington Carver*

Thrifting and Repurposing

The most sustainable way of doing things is to thrift when you can. If you spend less money, you can earn less money. This offers the freedom to work fewer hours off the farm and to live the home-based life you want to live. Do what you can to get away from your dependency on the mainstream market.

I wish we had realized earlier in our homesteading journey how much can be done by repurposing old materials. From purchasing clothing secondhand to reusing a dog kennel as the basis of a chicken coop and constructing raised beds from inexpensive cedar trimmings from a local lumber mill, so much of what we now acquire and build uses repurposed materials.

I love thrift shopping for clothing and home goods. When Miah's daughter Maliah is visiting, this is a favorite activity of ours. We call it treasure hunting. You never know what you can find secondhand, and you're getting it for a fraction of the cost of purchasing it new—not to mention cutting down on the environmental cost of manufacturing and transporting the raw materials and finished goods. Besides visiting local thrift stores, I buy a lot of my thrifted clothes through online auction sites such as eBay and online consignment and thrift stores such as thredUP, looking for brands that I already know will fit.

The same goes for materials on the farm and in the house. Have a general idea of the things you're looking for and start collecting them before you are in dire need. We were impatient when we first got started and ended up going into debt to build infrastructure and acquire equipment. This is not a sustainable way to go about it. Give yourself time to look out for sales and used items, even if you're not using them right away. Years into this lifestyle, we are still always looking for free stuff that we can use on the farm.

Browsing garden-supply catalogs, you may become convinced that getting started gardening costs a ton of money. You'll find self-watering container gardens for $500, raised-bed kits for $180, and more. These beautiful items

ABOVE: Cooking from scratch is a big step in resourcefulness.
OPPOSITE: A lot of food can be grown in repurposed containers.

are useful to some people, but if your homesteading budget doesn't stretch that way, you still have plenty of options.

Just because something is previously used or might not traditionally be what you would use in a project doesn't mean that it's worthless. Old privacy fence, reclaimed roofing tin, random lumber, wood pallets—these items are all easily reused and repurposed. The more skills you

Strawberries thrive for the second year in a row in an old cracked kiddie pool.

have with basic power tools, the easier it will be for you to build your homestead dream yourself.

Your ability to reuse and repurpose materials around the farm is only limited by your imagination. I once drilled holes in the bottom of a kiddie pool and repurposed it as a planter. We painted and hung tires around the garden on our first farm to grow flowers. You can even grow things right in the bag of soil that you bought, using the bag as the container.

It's important here to make a distinction between doing something in a thrift-minded way and doing it "cheap." Repurposing and finding good deals is cheap in the sense of being less money, but don't fall for cheap as in flimsy. Durable equipment and materials are a better investment than any poorly made cheap or free item that you'll have to spend time, energy, and money to replace in a few years.

Bartering

An energy exchange is another way to step out of the mainstream money economy. No one homesteads alone. You have skills that I don't have, I have skills that my neighbor doesn't have, and each of us exchanging labor and products makes our homesteads stronger.

We ended up with a herd of goats thanks to the bartering Miah did with a farmer, exchanging work for livestock. We would not have been able to afford this herd if we had to purchase these goats individually. These goats became a source of resources for our farm—milk and kids—that we have been able to further use for bartering.

You might have one afternoon a week that you can care for a neighbor's child in exchange for having weeds mowed in your pasture throughout the season. Perhaps the next time you split your beehive, you'll exchange a nuc for a livestock guardian dog. With bartering, anything is possible.

If ever there were an argument for building community, the ability to barter is a good one. You can trade and barter with people you don't know, but you're bound to have a better experience all around if you're working with friends and neighbors. There is an element of trust involved in bartering, knowing you're getting what's advertised.

Reusable Tattler lids mean not having to buy new lids every time we preserve our produce.

ABOVE: Chamomile reseeds every year, leading to an endless supply of blossoms for tea and medicinal salves.
OPPOSITE: The furled fronds of a fiddlehead fern taste similar to green beans and broccoli.

Foraging

Foraged food is found food. Learning to forage is a great skill for sustainability. I learned how to make jam using blackberries that I foraged on a friend's property. Well before I had a farm, that was my lifeline to this lifestyle. I loved picking gallons of blackberries and using this found fruit in every way imaginable.

Some of the sweetest foods—yes, berries—grow wild everywhere. In cities and towns across the more temperate climates, citrus trees and fig trees are abundant in public areas. Across the U.S. mid-South, pawpaw trees and elderberries galore are in surprising places with public access. Edible mushrooms and medicinal herbs, too, can be found if you know where to look.

Foraging is not an endeavor to be taken lightly. It's important to learn about the wild edibles that grow in your area before you set out on a foraging adventure. Respect that not all plants are edible. Many berries and many more mushrooms are toxic and can even be fatal. Some edible fungi even have toxic "lookalike" mushrooms. You must learn the difference. Get several reputable foraging guidebooks with photo identifications and go foraging with someone experienced before you attempt this on your own.

By starting out with someone who knows the ropes, you'll not only become comfortable with plant and mushroom identification but will also learn the best spots for foraging. Be sure you have permission to gather foods from a place before you dive in. Some cities have maps of public fruit trees, and out in the country, some landowners are known to be generous in granting access to their land. If someone allows you access to their property for foraging, say thank you with a jar of jam, pawpaw muffins, or elderberry syrup. I have found that a little bit of gratitude goes a long way.

It's easy to get carried away on a foraging adventure, filling baskets and buckets with your finds. You are sharing this wild food with the birds and wildlife as well as other people who happen upon it. Never harvest more than you can use, and try to keep your harvest to 30 percent or less of the bounty.

THE FIRST-TIME HOMESTEADER

Herbal teas, tinctures, and salves—both purchased and homemade—fill our medicine cabinet.

Homestead Medicine Cabinet

I believe herbs are meant to be one part of holistic wellness. It's still important to eat well, get enough sleep, drink your water, and seek self-care. In combination with a healthy lifestyle, herbs can support and nourish us.

I want to share some of the home remedies that we keep in our medicine cabinet—many of which have been used in various cultures for centuries. This is herbalism at its most basic. I am not a medical professional, nor am I an expert in herbalism. Nothing I say here constitutes medical advice.

Nature's remedies are everywhere. Many of the pharmaceuticals used in human and animal medicine today were originally or are still derived from plants. Valium (diazepam) came from compounds in valerian root, aspirin from willow tree bark, and morphine from poppies.

Herbalism and natural remedies are an overwhelming concept to many of us who are only now getting introduced to them. You may have questions about whether they work and are safe. Look to books about herbal solutions, and if you're not comfortable learning about these

subjects yourself, support a local herbalist who can guide you. There are affordable remedies, both store-bought and homemade, to get a basic medicine cabinet started.

I love growing herbs. Many herbs are perennials, so you plant them once and they keep growing over the years. Even the annuals are worth the trouble—not that they're any trouble at all. Herbs smell good in the garden, they attract pollinators and other beneficial insects, they're beautiful and interesting to look at, and they offer so many benefits after you've harvested them. (See "Herb Profiles" in this chapter for some of my favorite easy-to-grow herbs for tea and medicine.)

Using herbs fresh in food and medicine is a seasonal treat. Depending on where you live, you may be able to harvest from your perennial herbs year-round. Rosemary, oregano, thyme, sage, and others will keep producing through winter in mild climates. Elsewhere, even perennials go dormant in winter. This is where drying herbs comes in. Read my recommendations for drying herbs in chapter 8, "Homestead Kitchen Skills."

Making Tea

For accuracy, I'll note that the phrase *herbal tea* is incorrect. Tea comes from tea leaves (*Camellia sinensis*). A "tisane" is the actual name of what we're making here, with herbs instead of tea leaves steeped in hot water. Still, *herbal tea* is a common phrase, and this is what I call my garden beverage of choice.

In the heat of summer, a cold herbal tea is as refreshing as it gets. On a gray day in winter, there's nothing like a hot tea with honey. Many herbal teas are consumed for their medicinal benefits: Ginger or fennel for digestion, tulsi or lemon balm for a calming effect, and chamomile for sleep are a few examples of ways herbs may be used.

You'll learn the tastes of herbs over time and will come up with your favorite flavors in combination and alone. I like drinking hot chamomile tea with a little honey as well as a chamomile-tulsi-peppermint combo. Sage has a strong flavor and is refreshing when served cold and soothing when served hot. Lemon balm is delicious and lemony, both hot and cold.

You can make herbal teas with fresh or dried herbs:

• To make tea from fresh herbs, use 4 tablespoons (36 g) of fresh flowers or leaves (depending on the herb) steeped in 8 ounces (237 ml) of hot water. Cover the cup to keep the medicinal constituents that you'd otherwise lose in the steam. Bitter herbs like chamomile should only steep for 5 minutes; otherwise, they get very bitter. More aromatic herbs, like tulsi and lemon balm, can steep for 12 hours or more.

• To make tea from dried herbs, use 1 tablespoon (9 g) of dried herbs per 8 ounces (237 ml) of hot water. Cover and steep as you would using fresh herbs.

A warm cup of herbal tea in my favorite mug is my favorite way to calm my nerves and signal to my body it's time to rest.

Dried calendula, grown in my own garden, is a favorite to infuse with oil and use on irritated skin.

Making Syrups

If you enjoy a soda now and then, you may especially appreciate the simplicity and delight of herbal syrups. The kids call them honey medicine. These can be used for their medicinal benefits or simply for enjoyment.

Ginger tea sweetened with lemon balm syrup is a good tummy tea. Herbal syrup mixed with soda water is a refreshing drink. Black tea or lemonade sweetened with herbal syrup is a nice treat, too. My favorite herbal syrup uses lemon balm. Elderberry syrup is another that we use as an immune system support, one spoonful at a time as needed. Both recipes can be found in this chapter.

Plant medicine is not an exact science. When reading a recipe for an herbal syrup or other item that you'd like to add to your wellness cabinet, don't feel like you need every ingredient to make it worth it. Learn about the purpose of each ingredient so you understand what's going into your mixture, and do the best you can with what you have or with what flavors you prefer.

Use local, raw honey for your syrups whenever possible to take advantage of the natural benefits that it provides. You'll spend more money sourcing local, high-quality

honey (unless you're producing your own), but it's still a savings over what you'd pay for a finished herbal syrup from the natural foods store. (Be creative: Maybe you can offer a trade for the honey.)

LEMON BALM HONEY SYRUP

Ingredients:
Lemon balm
Water
Honey

Process:
Put your fresh or dried herbs in a pot and cover them with water. Heat over medium-low with the pot partially covered. Let the liquid reduce by half. This will only take 10 minutes for a small batch. Keep an eye on it.

Strain the mixture, and let it cool a bit. Mix two parts tea to one part honey. Keep it in a sealed jar or bottle in the fridge. Label it with the contents and date.

ELDERBERRY SYRUP

Ingredients:
¾ cup (109 g) dried elderberries
3 cinnamon sticks
2 Tbsp (12 g) shredded fresh ginger
1 tsp ground cloves
½ cup (64 g) dried rose hips
3½ cups (875 ml) water
1 cup (250 ml) raw honey

Process:
Put all ingredients except honey in a heavy-bottomed pot. Bring to a boil over medium-high heat. As soon as the boil begins, lower to a simmer and cover. Simmer for 45 minutes. Turn off the heat and let it cool to the point that it can be handled.

Remove the cinnamon sticks. Mash the remaining contents with the back of a spoon or potato masher. Press through a fine-mesh sieve, squeeze through cheesecloth, or run through a juicer. Mix 1 cup (250 ml) of honey into the resulting juice. Store in an airtight container in the refrigerator for up to a few months. Label it with the contents and date.

Elderberry syrup simmers on the stove before being strained and mixed with raw honey.

Fire cider is a staple through cold season. Start fermenting a batch several weeks before needed, then store in the fridge until needed.

Making Fire Cider

The name fire cider implies something hot—and that is accurate—but you are in control of its spice-intensity level. Its name comes from the idea that it warms up your body functions. At the onset of any cold-like symptoms, I take 1 to 2 ounces (30 to 60 ml) as needed.

Everyone makes fire cider using different ingredients, based on what they have on hand or what's in their grandmother's or herbalist's recipe. Garlic is a natural antimicrobial. Onion is an expectorant. Ginger and jalapeños stimulate your circulatory system. Lemon, lime, and rose hips offer vitamin C and a nice flavor. Turmeric is shown again and again to be anti-inflammatory. Together, they are meant to rev up the immune system.

FIRE CIDER

Ingredients:
½ cup (48 g) grated ginger root
1 medium onion, chopped
10 cloves garlic, chopped
2 jalapeño peppers
1 lemon, zested and juiced
1 lime, zested and juiced
Several fresh rosemary sprigs or 2 Tbsp (7 g) dried rosemary
Several fresh thyme sprigs or 1 Tbsp (3 g) dried thyme
1 Tbsp (8 g) turmeric
¼ tsp cayenne
¼–½ cup (32–64 g) rose hips
1 quart (1 L) apple cider vinegar
Raw honey to taste

Process:
Put all ingredients except honey in a large jar. Cover and place out of direct sunlight to ferment for 6 weeks. Use a vented fermenting lid, cover the jar with a tea towel secured with a rubber band, or open the lid to "burp" every couple of days so your jar doesn't explode. Stir or shake every few days.

After 6 weeks, strain. Return the liquid to a jar, and add honey to taste. Label it with the ingredients and date, and keep it in the refrigerator or root cellar.

Making Botanical Oil Infusions

Aside from herbs' usefulness in foods and beverages, they're great additives to skin care products. It's simple to make a botanical oil infusion and then to use that oil directly on your skin. You can also use it in your soap and salve making. Chamomile, calendula, and dandelion are especially good herbs for the skin. Lavender and rose add a nice fragrance.

One note about working with herbs and oil that isn't a concern when making tea: Use only dried herbs and flowers to infuse an oil. Fresh flowers and herbs contain too much moisture and can cause the oil to go rancid.

You want your infusion to last so you can use it over time. Oils with less saturated fat last longer because they are less likely to oxidize. Good oils for infusion are olive, almond, avocado, sunflower, grapeseed, and fractionated coconut. Jojoba oil (which is actually a liquid wax) is also a great option for topical use, as it is nourishing for the skin.

Simple oil infusions can be made with herbs for medicinal and culinary uses.

Chamomile oil can be used in place of plain oil in any skin salve recipe.

BOTANICAL OIL INFUSION

Ingredients:
Dried herbs
Shelf-stable oil

Process:
Loosely pack a clean, dry jar with fully dried herbs or dried fruits, leaving 2 inches (5 cm) of space at the top. Pour a shelf-stable oil over the herbs, covering with at least 1 inch (2.5 cm) of oil. Put on a well-fitted lid and shake. Place the jar in a sunny window. Shake daily for 3 weeks.

After 3 weeks, strain. Squeeze as much oil from the herbs as possible. Discard the strained herbs and put the oil in airtight jars. Label it with the contents and date, and store it in a cool, dark place. Use within 3 months.

Herbs for Animals

Herbs and natural remedies are not just for people. They can benefit our animals as well. In many parts of the world, animal health is still primarily treated using natural methods rather than pharmaceuticals.

Garlic is often looked to as a natural dewormer. I have used oregano and black walnut as a dewormer for the dogs, too. Broadleaf plantain, the common weed in your lawn, makes a natural poultice for sprains and bruises.

Herbs interact with animals' systems differently than with humans' systems. Catnip, for example, has a calming effect in people, but you know it to make your cats act boisterous. Saint-John's-wort—commonly used to elevate moods in people—causes photosensitivity in cattle, sheep, and horses. Some herbs are safe in small doses but toxic in larger quantities. It's essential that you understand plants' functions before offering them to your animals.

Herbs are useful for animal health, too.

Cleaning with Vinegar

Look in any homesteader's kitchen or pantry and you will likely find vinegar by the jugful. Vinegar plays a crucial role in pickling and preservation, but it's also an effective cleaner. With this one simple product, you can eliminate the need for many commercially produced cleaning products. This makes room in the budget and lightens the toxin load in your home.

You can use any number of vinegars to clean, but distilled white vinegar has the least residual odor and is also the cheapest. While vinegar cannot be called a disinfectant, acetic acid (its cleaning compound) is effective at reducing the bacteria *Escherichia coli* (*E. coli*) and salmonella, and it removes grime buildup.

Here are some ways I use white vinegar to clean:

• Dilute equal parts vinegar and water in a spray bottle for use on any food-contact surfaces (i.e., cleaning up spills in the fridge, on kitchen counters, and on baby's high chair).

• Spray vinegar on a greasy stovetop. Let it sit for 15 minutes, then wipe away the grime.

• Spray vinegar on cutting boards to clean.

• Mix equal parts vinegar and hot water in a spray bottle and use as a glass cleaner.

• For bathroom cleaning, spray vinegar in the sink, in and around the toilet, and in the bathtub and shower.

• For built-up grime, sprinkle baking soda first and then scrub with warm, undiluted vinegar on a scrub brush. (A baking soda–vinegar combo is the setup for the fun volcano experiment you probably remember from elementary school. Keep in mind that foaming action and go light on mixing these ingredients.)

• Freshen mattresses and fabric furniture: Mix 1 cup (250 ml) of vinegar and 20 drops of tea tree essential oil in a bottle and spray on the fabric. Allow it to air-dry. For lingering odors, spray and then sprinkle with baking soda. Vacuum after it's fully dry.

• Add 1 cup (250 ml) of vinegar to the final rinse of a laundry cycle to replace fabric softener.

• Remove pet odors by saturating the area with vinegar, rubbing in baking soda, and allowing it to fully dry before vacuuming.

We still use some store-bought cleaners, including antibacterial soap for cleaning our dairy animals' teats and udders, but vinegar is my go-to cleaning solution. Overall, making your own homemade cleaners is less expensive and makes for less waste, because you don't have to throw away all those spray bottles of individual cleaners purchased for each use in the home.

Resourceful living asks that you lessen what you need from outside your community. It's fewer lattes from the big-chain coffee shop and more homegrown tea with friends. It's fewer clothes-shopping trips to the mall and more thrift shopping—and still less shopping overall. It's a reliance on your home kitchen rather than the hottest new restaurants.

Reflect on your homestead goals. You're probably pursuing this lifestyle so you can step away from the rat race. That system is designed to keep you in it. Stay focused on your homesteading dreams so you can remove yourself from it. Know that you are creating a win with every small change you make toward a more self-sufficient life: every tool bartered for, every foraged-berry jam put by, every herbal remedy mastered. Making a lifestyle change is a journey, and you're on it.

TOP: *Every small step you take toward self-sufficiency for your family is a win.*

ABOVE: *I keep a large jar of dried elderberries so I can make a batch of syrup at the first sign of symptoms.*

Calendula (Calendula officinalis)

Chamomile (Matricaria chamomilla)

Herb Profiles

Calendula (Calendula officinalis)
If ever there were a cheery flower, calendula is it. With orange or yellow blossoms, its flowers make a nice fabric dye, a soothing topical oil, and an interesting salad topping. As its resin-y buds open, beneficial insects of all kinds make their way to this relative of daisies and marigolds. Flowers left unharvested produce the most interesting-looking seed pods in the garden. These are easy to collect for seed saving, or you can let the plant reseed itself.

To Grow: Start calendula seeds indoors 6 to 8 weeks before your last frost. Plant them just after your last frost date. Older plants can handle a light frost, but it'll shock the young plants. You can also direct-sow calendula seeds right before your last frost date. Calendula plants need regular watering and may suffer in the height of summer but can rebound as the weather cools off again.

To Harvest: Pick calendula flowers when they're fully open, just after the dew dries in the morning. Harvest continually to encourage the plant to grow bushy and make more blooms.

Chamomile (Matricaria chamomilla)
Chamomile's foliage is feathery and beautiful, and its delicate-looking yellow and white blooms are around all summer long. It's a nice plant to come home to. This herb has calming benefits. I like to drink it in a bedtime tea. The flavor is bitter if you over-steep the tea. It only needs to steep for 4 minutes. Add a little honey for a nice antianxiety beverage.

To Grow: Chamomile frequently reseeds itself, so plant it in an area where you won't mind enthusiastic chamomile volunteer plants in future years. To start, seed chamomile indoors 6 to 8 weeks before your last frost or seed it directly in the garden in early spring. The seeds are impossibly small and require light to germinate. Don't cover them with soil; just press them in. They have shallow roots and so require regular watering and care when being weeded around.

To Harvest: Ideally, you'll harvest early in the morning, when the petals are just opened and flat, which is when the flower has most of its beneficial properties. Harvesting a little before or a little after this point is also okay, but if the head is mushy, you've missed the window. Pinch off each head, one by one, leaving the rest of the plant to continue producing. The more you pick chamomile, the more flowers the plant will produce.

Comfrey (Symphytum officinale)
The botanical name of comfrey, *Symphytum*, means "unite." In common terms, comfrey has often been called "knit-bone" due to how it aids in the healing of broken bones. Comfrey is often controversial because of the risk of misuse. It is meant for external application only, as high internal doses can be damaging to the liver. It should also be avoided on serious lacerations, as it can heal the external skin too quickly, hindering the internal damage from healing properly. These cautions should not discourage you from growing and using comfrey, however. It may be extremely useful for sprains, bruises, inflammation, burns, and skin irritations.

To Grow: Comfrey is an easy plant to cultivate and is often seen growing wild. It is a perennial shrub that adapts well to most environments. It can be invasive and so should be planted in a contained area or an area where it will be welcome to spread. It can be difficult to start from seed and is most commonly propagated from root cuttings. Once you have a small piece of comfrey planted in your garden, you'll have plenty in no time.

To Harvest: Cut back comfrey branches to 2 inches (5 cm) above the soil for a large harvest or pluck individual leaves for a smaller harvest.

Elderberry (Sambucus canadensis)
This tree gives to us twice: first in elderflower form and then as elderberries. In the early summer, when the tiny, delicate elderflower blooms first open, pluck a cluster of them and add them to your water glass for a sweet, floral treat. Or leave the blooms and allow the berries to develop to make your own elderberry syrup and other immune system supports.

To Grow: You may have enough elderberry shrubs to forage in your area that you don't even have to plant any yourself. If you do want to grow your own, rooting cuttings is the simplest way to go. Put cuttings in a jar of water until they grow roots and then put them in a pot so they start to develop a root system. Keep them watered.

Comfrey (Symphytum officinale)

Elderberry (Sambucus canadensis)

Holy Basil (Ocimum tenuiflorum)

Plant them in the ground in the early spring or fall. Elderberry trees may grow up to 12 feet (3.5 m) in height, and you can keep yours pruned to make harvest more manageable.

To Harvest: Plan on 3 to 5 years for a new planting to mature. To harvest the elderflowers, wait until all buds in the cluster are open. It's best to get them in the morning on a less humid day. Every flower cluster you harvest is one fewer berry cluster available to you, so find a balance between the two. To harvest elderberries, have patience but not too much patience. There's a fine line between elderberries not being ripe enough and elderberries having been consumed by the birds. Look for a reddish-black to bluish-black color across all of the berries in the cluster. Cut the whole cluster and remove the berries one by one with care. Both flowers and berries are very perishable and should be processed right away.

Holy Basil (Ocimum tenuiflorum)
Adaptogens are herbs that help your body adjust to physical, emotional, and environmental stressors. Holy basil, also known as tulsi, may be the easiest of the adaptogens to grow at home. Delightful as a fresh or dried tea, in a fresh flower bouquet, and as a syrup for flavoring beverages, holy basil is an annual herb that always has a place in my garden.

To Grow: Start holy basil seeds indoors 5 to 8 weeks before the last frost, or sow directly in the garden when the threat of frost has passed. Holy basil will easily grow to 3 feet tall by 3 feet around (1 m by 1 m) if you don't keep up with harvest. Holy basil readily reseeds itself the following year, and you can collect the seed heads to save your own seed.

To Harvest: Holy basil is a member of the mint family, and the more you harvest it, the more it will grow. Tulsi stems are sturdy, so use scissors or garden snips for harvesting. Cut the center stem just above the growth nodes, where the plant branches. Those side branches will continue growing. Unlike culinary basil, holy basil has to flower before you harvest. You'll use both the flower and the leaf.

Lemon Balm (Melissa officinalis)

Lemon Balm (Melissa officinalis)
This perennial herb in the mint family is one of the easiest to grow, and because it's in the mint family, it's one of the most aggressive. I suggest planting this in a container or in an area where you don't plan to plant anything else because it will spread. Lucky for lemon balm, it's a favorite of mine: It has a delightful fragrance and flavor, and its antiseptic properties make it worth keeping around.

Lemon balm's bright green, scalloped leaves are vibrant from its first emergence in early spring until the first hard freeze of the winter. Lemon balm produces tiny white flowers that can almost go unnoticed by all but the bees that love them.

To Grow: Start seeds indoors 4 to 6 weeks before your last frost, or try direct-sowing in early fall. You can also divide lemon balm plants in early spring if you have a friend with an abundance. It should produce for three or more seasons, and then the oldest plants may need to be replaced or the newest seedlings allowed to take over.

To Harvest: Cut stems several inches from the top. Leaves are your primary target for drying and processing, but flowers won't hurt. Regular harvest encourages more leaf production and less seed production, which helps cut down on the self-seeding nature of the plant.

CHAPTER

8

Homestead Kitchen Skills

The kitchen is, undeniably, the heart of a homestead. All the sowing, reaping, and harvesting that takes place on the farm inevitably flows back to the counters of the kitchen where you, the homesteader, must decide how to best utilize the bounty.

Since taking on a role as a teacher of homesteading and gardening, I've learned there are two things people are quick to proclaim they are bad at: growing food and cooking it. It seems like anytime the topic of gardens or kitchen skills comes up, people quickly bemoan, "I burn everything! I kill everything! I can't grow a radish! I can't boil an egg!"

Darlings, this is nonsense. If you don't know how to cook, it is because you have not learned. Cooking does not need to be a complicated endeavor. What works for me in my kitchen may not be what works for you in your kitchen. What my family loves to eat may not be what your family loves to eat.

I intentionally choose to delight in cooking garden food. We like going out for dinner, but to be honest, the food is usually better at home because our ingredients are better. I know I've done a good job when Miah tells me he's more excited about what I'm cooking than about the possibility of going to any restaurant. It's taken some time to get here, and the reality is that a lot of my kitchen experience is trial and error. I don't get every meal right.

Everything you read in this chapter came out of this trial and error, and it is doable whether you live on a 50-acre (20 ha) farm or in a 500-square-foot (46.5 m²) apartment. Wherever you are on your homesteading journey, you can learn to cook and preserve food, to eat seasonally, and to get the most from the ingredients available to you.

"Until I discovered cooking, I was never really interested in anything."

—*Julia Child*

Tools for the Farmhouse Kitchen

The need for the right tool for the job applies to all areas of the homestead—the kitchen included. There are many gadgets and appliances out there that you may think you need, and you may be right about some of those. If you're like me and you have a small kitchen, you might be more interested in the basics of what it takes to process and cook with your own food at home. Following are the tools that I use every week, if not every day.

Cast Iron

I love cast-iron everything. When I moved into my very first home as an adult, the first thing I bought myself was a cast-iron pan, and now it is at the top of my list of must-have kitchen tools. I use my pans for baking, making bread, braising meats, sautéing vegetables, constructing casseroles, and making sauces—not to

My first cast-iron pan has been with me for two decades now, and I fully intend on having it forever.

mention, I can cook with them over a campfire or on a grill. Cast-iron pans are nearly indestructible.

Do not shy away from cast iron because of its reputation of being difficult to care for. It's a common misconception that you can't use soap on cast iron. You don't need to use soap because iron doesn't harbor bacteria, but if you do use soap, you just have to re-oil your pan when drying to maintain the seasoning (the slick coating that makes cast iron nonstick).

Iron doesn't heat up fast, but it does retain heat and cook items evenly. When seasoned properly, cast iron is very nonstick. Cast iron is healthier to cook with than chemical-laden nonstick varieties and even supplements the iron in your food. Cast iron is reactive, and some people say you shouldn't cook high-acid foods like tomatoes in cast iron because you'll get a metallic taste. I haven't had trouble with well-seasoned pans, though.

Enameled cast iron, like my Dutch ovens, is another option. Enameled cast iron is the same as regular cast iron, except you do use soap to clean it. Be aware that the enamel can chip over time in the lower-quality brands.

Cast-iron is heavy and it's important that it is kept fully dry. I like to store mine on a hanging pegboard.

A well-equipped kitchen makes using the food you grow so much more enjoyable.

Dehydrator

While you don't necessarily need a dehydrator for herbs, it's handy for dehydrating vegetables and fruits. Dehydration is a nice alternative preservation method, and it makes possible great foods like sun-dried tomatoes, fruit leather, veggie chips, and even beef jerky. Dehydrate tomatoes, grind them into a powder, and use that as you would tomato paste and to add a smoky depth of flavor to your meals. It's also great to have on hand a jar of dehydrated vegetables to put into a soup in winter. The vegetables reconstitute in the soup broth while it's cooking and take on so much flavor.

Dutch Oven

My enameled cast-iron Dutch oven is my second-favorite kitchen tool. You can find Dutch ovens in all sizes, enameled and not. I like my 8-quart (7.5 L) Dutch oven for cooking soups and stews. I also use these pots to bake, sauté, boil, simmer, fry, and roast all kinds of food. Being enameled, they're nonreactive, so I can use them for cheese making and yogurt making, too. These tools are heavy, so in choosing yours, find one that's the right size for your needs that is also not too cumbersome to handle.

Canning Supplies

From a canning pot to jars, find the basics covered later in this chapter.

Good Knives

It seems counterintuitive, but it's true that a dull knife is a dangerous knife. Sharp, durable knives are necessary for working with fresh vegetables and whole cuts of meat. You don't need a fancy chef's knife kit—just a few basic knives that you're comfortable using. In my knife block, I most often use:

• A chef's knife. These tend to be 8 or 10 inches (20.5 or 25.5 cm) long, and they're useful for chopping vegetables and doing most things in the kitchen.

• A paring knife is pretty much a small chef's knife, nice for small vegetables, peeling vegetables, and more precision cutting.

• A boning knife is, as the name implies, for cutting meat from bones.

• A bread knife is serrated and perfect for slicing home-made bread as well as tomatoes.

Good, sharp knives are a must for processing food.

Making a batch of butter in my Bosch mixer takes about 15 minutes.

Baking Sheets

You can find baking sheets in a range of sizes. I like to keep a variety of sizes so I can put multiple smaller sheets in the oven if I'm roasting several different things or just one large sheet if I'm roasting a lot of one thing. Baking sheets with lips are necessary for roasting anything that will give off a liquid, like tomatoes or squash, and also for corralling vegetables that want to roll off the pan as you stir them. Baking sheets without rims are nice for cookies (they're called cookie sheets) and pizzas. You'll find baking sheets in varying degrees of quality and in different materials. I like the stoneware baking sheets, though they're heavy. You can also invest in sturdy aluminum baking sheets.

Big Pots

This might go without saying, but you'll need some big pots to deal with the bounty of vegetables that come out of your garden. Big pots—12 quarts (11.5 L) and larger—are useful for canning, making batches of soups, blanching vegetables before freezing, making stock and bone broth (if you don't have a roaster oven), and more. It's nice to have a collection of cooking pot sizes so you can reach for the most appropriate size for your need.

Blender or Food Processor

There is a difference between a blender and a food processor, but if you don't have the space or budget for both, a food processor is the more versatile option. A blender is meant to crush and puree foods and requires liquid to do so. A food processor, with multiple attachments, can shred, chop, combine, puree, and more.

A convenient blender alternative is an immersion blender, also called a stick blender. This takes up less space because it's just the motorized handle and attachment, and while it's not as powerful as an actual blender, it can handle basic sauces, soups, and smoothies.

A Notebook

This tool is simple. I have a three-ring binder that is now a treasure, chockful of wisdom, notes, and recipes. While many people use Pinterest and save recipes in various places online, I like having something tangible that doesn't require computers or smartphones.

Roaster Oven

With our big family, I use our 22-quart (21 L) roaster oven often. It works both like an oven and a slow cooker. I use this mainly for making broth. It's also handy for making a second turkey or ham during the holidays and for rendering our own lard.

Though tools and gadgets are great, multipurpose items are also useful. I use a mason jar to cut out biscuits instead of having a specific tool for the job.

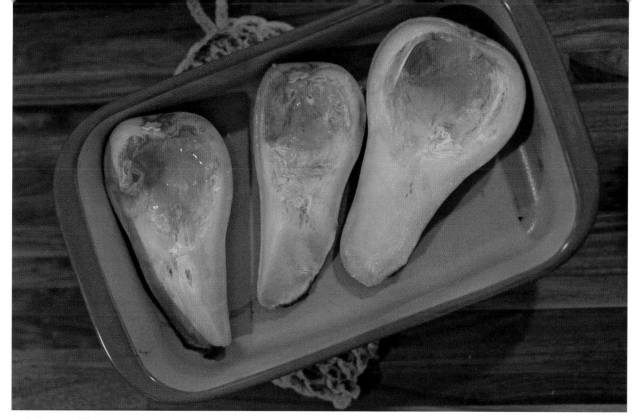

Lightly roasted butternut squash, which was grown in my friend Dina's garden.

Homestead Kitchen Know-How

I'm not great at sharing recipes because I don't cook with recipes. With good-quality ingredients, I like to keep things simple and let the food speak for itself. There are a few from-scratch cooking methods that I think a homesteader should become comfortable with. These are means of cooking that aren't so much recipes as they are skills. I'm sharing just a few of those with you here.

Roasting Winter Squash

Winter squash is a great garden crop because it keeps for months after harvest, and it's versatile. When you learn to easily roast squashes and pumpkins, you'll not want to buy the pumpkin puree in a can.

Wash your squash, cut it in half, scoop out the seeds, and place the halves face down on a baking tray. Roast in a 350°F (177°C) oven for about 40 minutes, until you can easily pierce the skin and flesh with a fork. Let the squash cool. Scoop out the flesh and run it through the food processor. Use this in soups, baked goods, and more.

Don't throw away those seeds. Roast them for a healthy snack. Rinse the seeds in a colander to remove any squash flesh. Lay them on a tea towel and dry them. Put them on a pan in one layer. Add 1 tablespoon (14 g) salted ghee or butter. Add seasonings if you'd like. Roast at 350°F (180°C) for 10 to 15 minutes.

Making bone broth in a roaster oven allows for a long slow simmer without tying up the stovetop.

Making Bone Broth

Chicken and beef bone broths are staples in my cooking. I use them as a base for soups, as my cooking liquid for rice, and more. I make big batches of broth at a time and freeze what I'm not using.

Put the stock bones in the roaster oven and roast on high to give them more flavor. Add water and a splash of apple cider vinegar and simmer on low for at least 24 hours, up to 48 hours. There is some debate about how long you should let broth simmer. Do some research and decide what's right for you.

If you don't have a roaster oven, you can use your regular oven to roast the bones and a slow cooker or a stockpot on the stove to make the broth. If you've already roasted the bones, as in making a whole chicken or bone-in beef roast, you can put those bones directly into the stock.

Roasting a Chicken

For years now, I've been roasting whole chickens using the spatchcock method, as this cooks faster than roasting a whole, intact chicken. If you were to roast the chicken without spatchcocking it, cooking will take about twice as long.

Use sharp kitchen shears to cut along one side of the backbone and press down on the chicken so it opens up and lies flat. Salt and pepper the skin heavily. Sear the chicken, skin side down, in a skillet with ghee or butter for about 4 minutes. Flip the chicken over in the pan and roast it in a preheated oven at 400°F (200°C) for about 45 minutes.

Peeling Tomatoes

This may be the largest time-saving technique I have learned for working with tomatoes. In canning and many recipes, peeling the fruit is part of the process. Those of you who've done this using the blanching-and-ice-bath method know that this is cumbersome.

Skip the boiling water and put your tomatoes in the freezer instead: Wash the tomato. Cut out the core and any cracked spots. Score the bottom (cut an X through the skin). Put the tomatoes in a freezer bag or container. Thaw before using. A lot of the juice runs out, and the skins slip off. Their texture is different after freezing, but the flavor remains. Use these in recipes or for canning.

A whole roast chicken takes over an hour to cook. Removing the backbone and pressing the chicken flat can shorten cooking time.

A glut of eggs led to perfecting the farmhouse quiche. This staple meal can be made with any variety of cheese, meat, or veggies.

Making Quiche

Quiche is a staple in our house come springtime and whenever we have eggs piling up. This is a no-recipe recipe. The basics are the same each time you make it, but it's up to you to add the seasonal fillings—whatever meat, cheese, and veggies you have on hand.

Start with a pie crust, either homemade or store-bought. In a bowl, beat 4 eggs, 1 cup (250 ml) of milk or cream, and your fillings. (Cook your meat and veggie fillings ahead of time, as they will heat through but not actually cook in the quiche.) Pour the egg-and-filling mixture into

the crust. Bake in a preheated 400°F (200°C) oven for 15 minutes. Lower the temperature to 375°F (190°C), and cook for another 25 minutes. When the top is golden brown, the quiche is done.

You can freeze a cooked quiche as well for a go-to meal. Let the quiche cool completely, wrap it in two layers of foil, and put it in the freezer. When it's time to eat, let it thaw in the fridge for 24 hours before reheating for breakfast, lunch, or dinner.

Preservation Time

I'm in my garden every day in early summer, checking every unripe tomato, asking, "Will you be the first one?" Come midsummer, I'm picking 30 pounds (13.5 kg) of tomatoes at a time. I have a dedicated kitchen day or two each week to process foods because this is a lot of harvest to keep up with. If I'm not being intentional with this time, stuff goes to waste.

The food you preserve will replace the cans and frozen packages of convenience foods that you would purchase at the store. Even better, you can control the ingredients that go into them. Between canning, freezing, pickling, fermenting, and dehydrating, you can put by food for your family to eat all year.

Canning

Opening your pantry and seeing row after row of neatly stacked jars filled with colorful vegetables that you put by yourself is a special thing. Canning is a time- and energy-intensive endeavor, yet I would not do without it.

Water-bath and pressure canning are the two different types of canning, and they are not interchangeable. Water-bath canning is the more basic method, used for high-acid foods, including most tomato-based products, fruits, relishes, vinegar pickles, some pie fillings, and the like. Water-bath canning does not reach as high a temperature as pressure canning. Pressure canning is for preserving salsa, green beans, pumpkin, meat, potatoes, and other foods without a lot of acid.

Items you need for canning include the following:

• A big pot: You can purchase an enameled canning pot specifically for this purpose, which is designed to hold many jars at once. You can also use a large stockpot in its place.

• Pressure canner: If you are pressure-canning foods, you need an actual pressure canner, not an electric pressure cooker. Purchasing new rather than used is smart, as you need to know it has a reliable seal.

• Jar lifter: This special set of tongs is handy for setting jars into and lifting jars out of the hot water in the canning pot.

TOP: *Pint-size Mason brand jars filled with local strawberry jam cool on the counter.*

ABOVE: *A jar grabber is a must-have tool for canning.*

THE FIRST-TIME HOMESTEADER

Pork rillettes cool in Weck brand jars.

Rillettes are potted meat covered with a cap of lard. They last for months in a fridge or a cool root cellar.

• Lid lifter: This plastic wand with a magnet on the end is used to put hot lids onto jars. It's nice but not necessary.

• Canning bubble remover tool: This is another nice-but-not-necessary tool, as a plastic knife or small spatula can do the same job. The actual tool does have useful markers on it to measure jar headspace, though.

• Wide-mouth funnel: These funnels make it easier to get the ingredients into your jar instead of all over your counter.

• Jars, lids, and bands: These are covered below.

Be prepared for canning season well ahead of canning season. Start collecting your jars, lids, and bands before everyone else starts collecting theirs, meaning don't wait until July to purchase these items. I'll buy a case of jars when I'm at the store outside of canning season because jars are easier to find then, and if I buy one case at a time, I'm spending an extra $10 or $15 on jars that week, not $100 or $150 on jars all at once. Sometimes you'll even find jars and lids on sale.

Invest in good-quality rather than inexpensive off-brand jars. Jar breakage is a regular part of canning, but you'll have a lot more breakage and many more seal failures if you go cheap. Ball and Kerr jars are the way to go.

Canning jars come in various sizes, from ½ cup (125 ml) to ½ gallon (2 L), and in regular mouth and wide mouth.

Every regular-mouth jar and every wide-mouth jar uses that-sized lid and band, no matter the volume or the brand of the jar.

Know the difference between a canning lid and a storage lid. To preserve the contents of a jar, you need a new metal lid and a metal band. The band can be reused, but the lid must be new to ensure a good seal.

The full canning process depends on the vegetable, meat, or recipe you're preserving. Basically, you process your ingredients and fill the jars with the prescribed amount of headspace between the food and the top of the jar. You wipe off the rims of the jars, place the lids on top, tighten the bands, and process the jars in either a pressure canner or a water-bath canner. The heat creates a vacuum in the headspace and removes the oxygen from the jar. In the process of boiling the jars for the prescribed amount of time, an internal temperature is reached that kills contaminants that might exist inside.

I am hesitant to give advice on proper canning recipes because of the food-safety nuances involved. There are a lot of recipe and food-safety resources available from the National Center for Home Food Preservation, your cooperative extension or local governmental agricultural service, and canning-jar manufacturers.

Pickled vegetables can be water-bath canned to be shelf stable or they can be stored in the refrigerator for an extended period if you prefer not to can them.

Quick Pickling

An easy, quick—like the name—way to preserve vegetables is quick pickling. Also called refrigerator pickles, these instantly gratifying pickled snacks and side dishes can last in your fridge for weeks or more, but you'll probably eat them before then.

You'll find all kinds of recipes for quick pickles, because just about any combination of vegetables, spices, and herbs goes. The central idea is the same: Heat a brine of equal parts vinegar and water, ½ teaspoon of salt per cup (250 ml) of liquid, and some combination of sugar, herbs, and spices. Pour the brine over raw vegetables in a clean jar or bowl. Allow this mixture to cool to room temperature and then put it in the fridge for flavors to meld.

Quick pickles are not meant to be canned—just stored in the fridge. Because you're not running a canner, small batches of quick pickles are just as worthwhile as large batches. If you're not sure what to do with a handful of end-of-season cucumbers, that cauliflower you forgot to cook with dinner, or a few onions you need to use before they go bad, quick pickling is an option.

Freezing

Freezing is an easy preservation method, though it is energy intensive, both in that it requires some processing to get foods ready for freezing and that running a freezer requires electricity.

Most vegetables can be frozen, though they do not retain their texture once thawed. Some vegetables require blanching (cooking briefly in boiling water and then immersing in ice water to stop the cooking action), steaming, or boiling before freezing. These include greens, green beans, okra, broccoli, asparagus, and corn. Others, like chopped onions and diced winter squash, can be frozen as is.

Preserving food by freezing isn't limited to vegetables. Making a big pot of a filling soup is great if you're feeding a houseful and if you want to stock your freezer with meals that are ready to go. Broths freeze beautifully. You can also freeze roasted and shredded chicken in the portions you need for your favorite recipes to save you time later. Your freezer can be a great source of convenience ingredients like this.

Multitasking is a big part of kitchen preservation days. On this day, I canned twenty-four jars of strawberry jam while also starting a large crock of sauerkraut.

Fermenting

Kimchi, sauerkraut, and pickles are all examples of fermented foods. Kombucha, kefir, and sourdough breads are fermented as well, and while we love those foods and drinks, I am not able to cover them here. Like quick pickling, the fermented vegetable combinations you can create are unlimited. Besides being tangy and delicious, fermented foods introduce probiotics, good bacteria, and enzymes into your diet.

Different ingredients require slightly different fermentation conditions, based mostly on the water content of the vegetable. Some require a brine of water and salt, and others, like cabbage, require salt only and will give off enough liquid to create their own brine. Each needs a bit of time at room temperature for the lactobacilli bacteria (a good bacteria) to go to work preserving your food. Abundant fermentation recipes and advice exist in books and online.

Preserving Dairy

If you're getting into dairy animals, learn how to use excess milk. Yogurt, butter, kefir, and cheeses keep longer than fresh milk, not to mention they are delicious and probably part of your kitchen staples.

With both of our cows in milk as I write this, we're getting 4½ gallons (17 L) of milk every morning. That's a lot, and even with a family of our size, we could not possibly drink it all. I've sharpened my cheese-making routine with fresh mozzarella and fresh ricotta, as well as become more serious about separating out the cream and making butter. The ability to make butter, you might know, was a primary motivation behind bringing cows to our farm.

When making cheese, the fresher the milk, the better. If using store-bought milk, look for the kind that's not ultra-pasteurized or shelf stable.

Cheese making can be as complicated an endeavor as you'd like to make it. Aged hard cheeses ask for special molds or presses and ingredients that can be hard to source. To make simple, fresh cheese, you only need milk, an acid to cause the whey to separate from the curds, and salt. With a little effort, you can make yet another food that you won't have to buy from the grocery store.

Butter made from cream in the spring, when the grass is growing richly, is a deep golden color.

Herb Preservation

Something about bundles of herbs drying in the kitchen just makes me feel like I'm in a farmhouse. Growing and preserving your own herbs is a great way to offset the grocery budget and take your cooking to the next level. I like using herbs fresh whenever possible, but home-grown and home-preserved herbs are still worlds above the commercially produced products available at the store. Herb plants often grow large and benefit from pruning, and you can use this harvest to stock your pantry, not to mention that dried herbs make great gifts.

Drying Herbs

To keep herbs on hand for cooking, making oil infusions, and teas, drying is your best bet.

Harvest your herbs early in the morning, after the morning dew is dry but before the heat of the day.

If you choose to rinse your herbs before drying, use cool water and then gently shake off excess moisture. Remove any damaged leaves or debris and lay the herbs in a single layer on a dish towel. Throughout the day, turn the herbs so they are fully dry of all surface moisture.

Choose from several drying methods:

• To air-dry: This is my favorite method for drying herbs, as it is the least damaging to the naturally occurring oils and therefore preserves the highest flavor and health benefit. Simply tie the herbs into small bundles (no thicker than 1 inch [2.5 cm] to avoid mildew forming in the center) with twine or a rubber band. Hang upside down out of direct sunlight in an area with good airflow. Depending on the moisture content of the herbs, they may be fully dry within a week but can take up to a few weeks.

• To dry in a dehydrator: A dehydrator or an oven can be used to dry herbs, though use care, as overheating herbs can greatly detract from their flavor and medicinal properties. You might prefer this method during damp or cool times of year when air-drying will take longer. Lay herbs in a single layer on dehydrator trays. Dry on the lowest setting of your dehydrator. Start with 3 hours of drying and check for doneness. Continue in 1-hour intervals until fully dry.

Herbs grow abundantly and add luxury to our food.

• To dry in an oven: Set the oven on the lowest setting. This should be lower than 200°F (95°C). The "keep warm" setting is preferable if your oven has one. Remove larger herb leaves from stems or keep intact if leaves are smaller. Lay herbs in a single layer on a baking sheet. Bake for 30 minutes with the oven door cracked open to let moisture escape. If your oven cooks unevenly, rotate the trays. After the first 30 minutes, check dryness. Continue in 10-minute intervals until fully dry. Cool completely before storing.

Herbs are fully dry when they crumble easily between your thumb and forefinger. If any part of the leaf still rolls or feels gummy, it's not dry enough for storage. They lose potency with air exposure, so keep dried herbs in airtight containers out of direct sunlight. Avoid crumbling fine or grinding until using. They have their best flavor when used within a year.

Use pipe cleaners from the craft store to easily hang herbs to dry. These are bay leaves I received from a friend.

Freezing Herbs

Herbs' texture will be ruined by freezing, but they maintain most of their flavor for cooking. Herbs frozen in water can be added to pots of soup and broth. Herbs frozen in oil can be added directly to a skillet and sautéed.

After harvesting and rinsing, rough chop your herbs and pack them into ice cube trays or molds. I prefer silicone trays for easier removal. Cover herbs with water or oil and freeze until solid. Remove herb cubes from trays and store them in a labeled plastic bag in the freezer for up to 6 months.

I want you to feel empowered to use what you grow and make choices that are sustainable for you. Providing food for your family means doing your own research and learning. There are different ways to do nearly everything in the kitchen, each carrying its own level of skill and food safety. Decide what you are comfortable with, what you think is best, and what you want to create for and give to your family.

Any time spent in your kitchen, learning new kitchen skills, putting by food, and preparing meals is time well spent. As you look into your pantry, freezer, and refrigerator—and look less to a grocery store—for your meal options, the full homesteading picture comes into view. Here, you are creating real, unadulterated food that's as local as you can ever hope for.

A single loaf of artisan bread can cost several dollars at the store versus the homemade cost of less than a dollar. Scratch cooking is healthier and far more affordable.

Cooking from scratch is labor of love, and it's how I express my heart to my family.

Top Homestead Kitchen Tips

Learn to cook from scratch.
Knowing what to do with the food you grow is a big part of success in producing your own food. You're going to be more enthusiastic about your gardening, chicken keeping, and home dairying when your family is enjoying the bounty from them. Cooking from scratch is significantly less expensive than going to restaurants and often less expensive than buying packaged food from the grocery store. You know where your ingredients come from, and you can eat in alignment with your values.

I encourage you to take risks and try new things in your homestead kitchen. Try your hand at cooking without a recipe. I'm not suggesting that you bake without a recipe, because baking requires exact science. Don't try canning something without a tested, food-safe recipe either. Start with a dish composed of basic ingredients, maybe one you've made before. The easiest thing to cook without a recipe is vegetables, straight out of the garden. Try a summer-vegetable stir-fry with fresh herbs. Roast root vegetables with different combinations of pantry seasonings. Make a soup with sausage, beans, and a combo of vegetables. You'll get the hang of it and increase your confidence.

Eat Seasonally

It may seem like common sense that if you're growing your own food, you're going to eat it in season, but for most of us being so far removed from our food sources, this can take a lot of thought. When you get on the farm and you have a flock of chickens and you find you have 6 dozen eggs per week, you begin to realize it's time to eat more eggs: quiches, omelets, egg salad, fried eggs on top of everything. The same goes for asparagus in the spring, zucchini in the summer, kale in the fall, and sweet potatoes in the winter. Eating seasonally means eating flexibly. You will gain an appreciation for each food as it comes and goes with the change in season.

Eat leftovers.
When you grow your own food, you don't want to waste any of it. Because we often have extra people over and we have hungry teenagers who are always ready for a hardy snack in the afternoon, we keep a lot of leftovers.

In addition to eating leftovers in their original form, you'll become more creative with turning your leftovers into new meals. Potatoes and vegetables you roasted for dinner last night become fillings for a quiche this morning, leftover roasted chicken makes a great topping for a salad, and leftover beans are the start to a soup with garden vegetables.

Let vegetables be the star.

When you become serious about gardening, you end up with many more vegetables than any other food group. It's unusual for us to have a dinner that is a hunk of meat on the plate. Occasionally we'll do something like burgers, but with a family of eight, you're not going to sit everyone down to a dinner of their own chicken breast. The meat from one chicken feeds the whole family. We eat a lot of soups, stews, and bean-centered meals, all full of vegetables. These are things we add meat to as a seasoning or as a side.

Stock your pantry.

In my pantry, I like to have a supply of food that can feed my family for a couple of months. You can learn to shop sales so you can pick up three or four of the on-sale items when your budget allows. Be organized with your pantry, and rotate foods as they come in so you use the oldest first. This backup allows you to stretch meals when your budget is tight, the supply chain is stressed, or your whole family gets sick and you just can't get to the grocery store.

Buying pantry items in bulk is smart, when possible. Dried beans, grains, pasta, and other shelf-stable whole foods can keep for a long time, and it's nice to reach into the pantry instead of driving to the grocery store every time you want to make a favorite recipe. Of course, filling your pantry with foods you've grown and preserved yourself is the best pantry-stocking tip of all.

Dried herbs and vegetables grown in the garden can be enjoyed year-round.

Conclusion: A Simple and Beautiful Life

When I began this homesteading dream, I didn't know where it would take me. I trusted that the journey would be worthwhile, and to say it has would be an understatement. This journey has allowed me to live closer to the land, seek sustainability, and grow food for my family. It has not always been easy, as you've read in these pages, but it has always been worth doing.

For me, this homesteading path began with a review of my values and where I fit in to consumer culture. I started to learn how to cook in a small kitchen using store-bought and some farmers' market food, then I learned to grow food for my family in rocky, less-than-ideal soil, took on too many chickens before I was ready, and learned all about dairy goats. Now we're building the homestead of

my dreams, complete with enthusiastic children, a few dairy cows, and bountiful, beautiful gardens. Looking ahead to those I've learned from along the way and looking back to help those coming in behind me, I know this community is where I belong. There is plenty of space here for everyone who wants to join—you included. I believe there has never been a better time to start homesteading or to start making plans for homesteading.

When we first got into homesteading, it was coupled with so much fear. This was my dream come true: to live more sustainably and simply. I was enamored with this beautiful, romantic life. At the time that we started to realize our dream, we were also realizing the climate of the modern-day world, and a fear was bubbling. If you are experiencing this same feeling, keep your heart in check. Consumed by fear, you'll make decisions that are not in your best interest but in your fear's interest. Wisdom, from the heart, is where we need to operate. Hold on to why you have this homestead dream, and let wisdom guide you into it.

Even if you're still in your waiting room, gain skills now and make this time part of your homestead journey. Learn and research. Read, watch other people, and hypothesize about how you'll do things. Know that a lot of things just have to be learned experientially, despite your best knowledge and planning.

Use your homesteading dreams as a tool to guide your steps in the present. Partner with your dream and bring it into reality, step by step.

It's hard to be intentional, to not live in your head months into the future and long for what's next. I want to live in the moment and enjoy the season that I'm in. I mean that in the literal sense, as in appreciating the cold of winter as a time of planning and restoration, and in the larger sense, as in working with and appreciating the circumstances I have now instead of always wanting what's to come.

Taking a walk around the farm during the evening golden hour and seeing the spaces that we've built is a beautiful way to end my day. There is a richness available to us when we choose to cultivate gratitude. I still sit in awe and wonder that this golden-hour farm time, these hilarious chickens, the gorgeous dairy cows, this very life has happened for me. I did not start with this farm-wide view. No part of the homesteading lifestyle I am living now came without planning, work, patience, and trust. Even though I might throw up my hands at the animals and want to quit the garden sometimes, I return to this work each day because it's the result of a dream that I have stuck with for almost as long as I can remember.

There is so much value in pouring your heart into something and creating something that moves you. I know that you can do this.

Until next time, I bless you.

Jessica

BELOW: Top row (left to right): Asher, Jackson, Jeremiah, Jess, Maliah. Bottom row (left to right): Ezra, Jean Luc (the chicken), Benjamin, and Tobias.

About the Author

Jessica Sowards is a native Arkansan now living in South Carolina who is homesteading, gardening, and sharing her journey through social media. She and her husband, Jeremiah, are raising their six children on a small hobby farm. They share their lives on their YouTube channel, blog, and social media platforms, Roots and Refuge Farm. Jessica is passionate about gardening, growing food, raising animals, and equipping others to do the same.

Acknowledgments

An acknowledgments page seems a woeful attempt to show the proper appreciation of the massive group effort that goes into the creation of a book. The simple fact that I could ever have a book on homesteading to my credit is a humbling truth that leaves me in a place of great gratitude and praise. It is infinitely more than I even thought to hope for when I used to long for a big red barn and chickens in the yard. Thank you to my Sweet Miah for laboring so diligently in seeing my imagination surpassed and, of course, to my Abba for making it all possible.

To my brilliant children, I pray the far reach of your mother's wildest dreams will embolden you to be crazy brave in your lives. You are able and worthy, and my greatest prayer is that my ceiling becomes your floor and that you know how deeply you are loved.

To my parents, thank you for convincing me that I could do anything. I really do believe you.

To our YouTube subscriber friends, your faithfulness and support make it possible for me to hold the homestead door open for others. I cannot express what an honor it is to get to do that. Thank you.

Papa T, Mimi, Erin, and Mike, you made our transition from Arkansas to South Carolina one of joy instead of struggle. You were a balm during the process of writing this book and played a huge part in establishing home. Thank you.

Kathy, thank you for the encouragement you spoke over the importance of these teaching books. Amy, thank you for always being at the ready with a pep talk and for the photo of me on the cover. Thank you, Justin and Rebekah, for letting me photograph your cows before I had my own and for all the inspiration you've shared. To Wes and Laurie, thank you for letting me photograph your rabbits and for how fully you believe in our big vision.

To everyone at Cool Springs Press and Quarto for your work, and specifically to Jessica, Heather, David, Steve, and Lisa for everything you poured into this book. It is great because of you.

And to every reader that consumed the words of these pages, my earnest hope is that your life is better for it. I pray your homestead dreams come true in ways that are wilder than you ever hoped, and I pray you know the joy of a simple life and a plate of food you grew yourself.

Index

A

A2 milk, 102

Alpine goats, 93, 96

Ameraucanas (chickens), 38, 45

American Milking Devon cows, 97, 99

Animals. *See also* specific animals
    butchering at home *vs.* outsourcing, 72–73
    for dairying, 78–80, 81
    feeding. *See* Feed(s) and feeding
    herbs for, 160
    housing for. *See* Housing
    industrial meat production and, 52
    symbiosis with the garden, 123
    veterinary care of, 72

Annual gardens, 108

Antibiotic-resistance bacteria, 52

Araucanas (chickens), 45

Artificial insemination, 65, 67, 89

Asparagus, 108, 109, 182

Automatic coop doors, 44

B

Baking sheets, 172

Barred Rocks (chickens), 37

Bartering, 151

Beans, 115, 119, 124, 126, 178

Bedding, chicken coop, 43

Beef, raising, 66–71

Bee keeping and bees, 129–145. *See also* Hives
    gear for, 135–136
    hives, types of, 130–132
    native, 143–144
    natural *vs.* conventional management of, 141
    protecting hives and, 142
    secret lives of, 132–133
    siting a yard for, 140
    sourcing, 136–139
    sourcing bees for, 136–139

Bee suit, 135

Belted Galloway cattle, 71

Berkshires (pigs), 65

Berries, 108, 109–110

Bielef elders (chickens), 38

Birthing and birthing kit, 84, 91

Birthing stall, 84

Black Angus cattle, 71

Blackberries, 108, 109–110

Black Copper Marans (chickens), 37–38

Black Rangers (chickens), 55

Blender, 172

Blueberries, 110

Bone broth, making, 174

Boning knives, 171

Botanical oil infusion, 159

Bottle feeding, 92

Bread knives, 171

Breeding
    cows, 67, 90
    goats, 90
    home dairy and, 89, 90
    pigs, 65
    poultry, 57
    rabbits, 60–61

Breeds
    cattle, 71
    chicken, 36–38, 45
    dairy cow, 97–99
    dairy goat, 93–96
    duck, 48
    geese, 48
    pig, 65
    poultry, 54–55
    rabbit, 61

Breeds, chicken, 36–38

Broad-Breasted White, 54

Brood boxes, 130

Broodiness, chicken, 36–37

Brown Swiss cows, 97, 99

Bucklings (baby goats), 90

Bucks, 89

Bulls, 67, 89

Butchering animals, 72–73

Butterfat, 93, 100–101

C

Calendula, 159, 162

Californians (rabbits), 61

Canning/canning supplies, 170, 176–177

Carrots, 112, 124

Cast iron pans, 168

Chamomile, 155, 159, 162–163

Champagne d'Argent rabbits, 61

Cheese making, 179

Chef's knives, 171

Chicken(s), 12, 31–49. *See also* Hens
    after egg production ends, 42
    breeds of, 36–38
    building a flock of, 32–35
    chicks, purchasing, 34–35
    coops for, 40–41
    egg production, 32, 45–46
    eggshell color and, 44–45
    feeding, 42, 45
    hens, purchasing, 34
    housing, 56
    for meat, 54–55
    for meat production, 57
    protecting from predators, 41, 44

THE FIRST-TIME HOMESTEADER

reasons for low production of, 45–46

roasting a, 174

roosters, 38

using manure from, 43

"Chicken math," 31

Chicken tractors, 40

Chicks, 34–35, 36, 40, 55, 57

Chickshaws, 40

Chores, 11–12

Cleaning, 160–161

Clothing, thrifted, 148

Cochins, 37, 38

Cold composting, 120

Colony Collapse Disorder (CCD), 144

Comfrey, 163

Compost(ing), 43, 120–121

Concentrated animal feeding operations (CAFOs), 52

Cooking

pantry for, 183

from scratch, 182

Coops, chicken, 40–41, 44

Corn, 124, 126, 178

Cornish Crosses (chickens), 54

Cows, 11–12, 66–71

birthing and, 91

bottle feeding and dam raising, 92

breeding, 67, 90

breeds of, 97–99

daily average production of, 99

for dairy, 78, 81

feeding, 88

fencing for dairy, 83

milking, 86–87

personality of, 101

Crested Cream Legbars, 45

Cuckoo. See Marans (chickens)

Cucumbers, 114, 115, 126

Currants, 110

D

Dairy. See Home dairying

Dam raising, 92

Dandelion, 159

Dehydrators, 170, 180

Delawares (chickens), 37

Devon cattle, 71

The Doe Code, 91

Doelings (baby goats), 90

Does (cows), 90

Does (rabbit), 58

Drainage, land shopping and, 20

Drying herbs, 154, 180

Ducks, 47–48

Dutch oven, 170

E

Eastern Bobwhite quail, 46–47

Egg mobiles, 40

Eggs, determining number of chicken based on production of, 32

Eggshell color, 44–45

Elderberry, 163–164

Elderberry Syrup, 157

Electric wire/electric net fencing, 65, 69, 83

F

Fabric mulch, 122

Feed(s) and feeding

bottle, 92

chicken, 42, 45

cows, 66–67, 88

medicated, industrial meat and, 52

milking animals, 88

pigs, 64

sourcing, for animal agriculture, 53

Fencing

for cattle, 69

for geese, 49

for goats, 83

for pigs, 64–65

for poultry, 41, 56

Fennel, 155

Fermenting foods, 179

Fertilizer, 100, 119

Fire cider, 158

5-year plan, 28

Food preservation, 176–179

Food processor, 172

Foraging, 64, 152

Freezing foods, 178

Freezing herbs, 181

Fruit trees, 106, 108

G

Gardens and gardening, 105–126

building soil for, 118–122

farm animals and, 123

kitchen (annual) garden, 108

perennial food garden, 108–110

planning for, 111–114

potager, 116–117

site for, 106

succession planting in, 114–115

year-round growing in the, 115

Geese, 48–49

Ginger, 155, 157, 158

Ginger Broilers (chickens), 55

Gloucestershire Old Spots (pigs), 65

Goats, 12, 81

birthing and, 91

bottle feeding and dam raising, 92

breeding, 90

breeds of, 93, 95–96

daily average production of, 96

for dairy, 78, 80

feeding, 88

fencing for, 83

kept in the suburbs, 18

milking, 86–87

personality of, 101

Grain-fed beef, 66
Grass clippings, 122
Grass-fed beef, 66
Grazing, 66, 82, 97
Guernsey cows, 98, 99, 102
Guineas, 46

H

Hampshires (pigs), 65
Hand milking, 85
Hatcheries, buying chicks from,
    34–35
Heifer, 90
Hens
    after egg production ends, 42
    broodiness, 36–37
    purchasing, 34
    in the suburbs, 18
Herbalism, 154–158
Herbs
    for animals, 160
    botanical oil infusions made
      with, 159
    drying, 180
    freezing, 181
    in the perennial garden, 110
    profiles of, 162–165
    succession planting, 115
    syrups made with, 156–157
    teas made with, 155
Herefords (cows), 71
Heritage breeds of cattle, 71
Heritage chickens, 36–37
Heritage poultry breeds, 55
Highland cattle, 71
Hives. *See also* Bee keeping and
    bees
    aggressive, 145
    inspecting, 141
    placement of, 140
    protecting, 142
    types of, 130–132
Hive smokers, 135
Hive tool, 136

Holstein cows, 88, 98, 99, 102
Holy basil, 164
Home dairying
    birthing and, 91
    bottle feeding and, 92
    breeding and, 89, 90
    butterfat and, 100–101
    dam raising and, 92
    feeding milking animals, 88
    fencing and, 83
    goats *versus* cows for, 78–80
    housing for, 84
    mentorship for, 81
    milking and, 86–87
    milking room for, 85
    pasture for, 82–83
    preserving dairy, 179
    uses of milk, 100
Homeowners associations, 18, 20
Homesteader(s)
    choice and determination for
      being a, 14–16
    different types of, 14
Homesteader kitchen, 167–183
    canning and, 176
    cooking and, 182–183
    preserving foods, 176–179
    skills for, 173–175
    tools for, 168–172
Homesteading
    author's personal experience
      with, 8–9, 14–16
    challenges with, 21, 23
    daily activities and chores
      with, 11–12
    definition, 14
    growth in popularity of, 9
    journey of, 184–185
    knowing your why for, 28,
      236–237
Homestead layout, 24–26
Honey, 132, 133, 141, 143
    Elderberry Syrup, 157
    Lemon Balm Honey Syrup, 156

Honeycomb, 132
"Hot" composting method, 121
Housing
    cattle, 69
    dairy animals, 84
    ducks, 48
    pigs, 64–65
    poultry, 56, 73
    rabbits, 58
Hybrid poultry, 36–37, 54–55

I

Industrial meat, 52

J

Jersey cows, 88, 97, 98, 99, 102
Jerusalem artichokes (sunchokes),
    110
Jumbo Bobwhite quail, 46–47

K

Kitchen gardens, 108, 116
Kits (baby rabbits), 58–59
Knives, kitchen, 171

L

Lamanche goats, 93, 96, 101
Land
    laying out the, 24–26
    purchasing raw, 20
    shopping for, 19–20
Langstroth hives, 130
Large Blacks (pigs), 65
Laws, land shopping and, 19–20
Leftovers, eating, 182
Lemon balm, 110, 155, 165
Lemon balm honey syrup, 156
Light, egg production and, 45
Livestock Conservancy, 35

M

Mail-order, purchasing chicks
    through, 35
Mangalitsas (pigs), 65

Marans (chickens), 37–38, 45
Meat, 51–75
    feed sourcing and, 53
    home butchering vs.
      outsourcing, 72–73
    industrial, 52
    packaging and storing, 75
    from poultry, 54–57
    from raising cattle, 66–71
    from raising pigs, 62–65
    from raising rabbits, 58–61
    storing, 75
Medicinal herbs, 154–157
Mentorship, 65, 81, 129, 137
Milk
    A2, 102
    goat versus cow, 80
    pasteurized, 102
    raw, 102
Milking, 11–12, 77, 86–87
Milking room, 85
Milk machine, 85
Miniature beef cattle, 71
Mint, 110
Mites, 45, 141
Mobile chicken coops, 40, 43
Mobile rabbit housing, 58
Modern homesteading, definition
    of, 14. See also Homesteading
Molting, 46
Mulberries, 110
Mulch(ing), 119, 122
Murray McMuarry Hatchery, Iowa,
    35
Mushrooms, foraging for, 152

N

National Poultry Improvement
    Program, 34
Native bees, 143–144
Natural remedies, 154
Nectar, 133
Neighbors, 20, 140

New Zealand White rabbits, 61
Nigerian Dwarf goats, 90, 93, 96, 101
Nubian goats, 95, 96, 101
Nuc (bee colony), 136

O

Oberhasli goats, 95, 96, 101
Orpingtons (chickens), 37, 38
Oyster shell, for chicken, 42

P

Package (bee), 136
Packaging meat, 75
Pantry items, 183
Paring knives, 171
Pasteurized milk, 102
Pasture
    cattle and, 71
    dairy animals and, 88
    for geese, 49
    industrial meat vs. animals
      kept on, 52
    pigs and, 62, 64, 65
    poultry and, 55, 56
    setting up the dairy and,
      82–83, 88
Penedesencas (chickens), 45
Peppers, 112, 124, 126
Perennial food garden, 108–110
Permaculture, 25, 116
Permaculture zones, 25–26, 106, 140
Pickling, 178
Pigs, 12, 62–65
Pineywoods (cows), 71
Plymouth Rocks (chickens), 37
Pollen, 133
Potager garden, 116–117
Pots, 172
Poultry. See also Chicken(s)
    breeding, 57
    breeds of, 54–55
    housing, 56
Poults, 56, 57

Predators, 41, 44, 56, 58
Preservation
    dairy, 179
    dehydrators for, 170
    food, 176–179
    herbs, 180–181
Processing animals, 72–73
Propolis, 133
Pumpkins, 126, 173, 176

Q

Quail, 46–47
Quiche, 175

R

Rabbits, 58–61
Raspberries, 109
Raw land, 20
Raw milk, 102
Reclaimed and repurposed
    materials, 148–150
Red Angus cattle, 71
Red Rangers, 55
Resourceful living, 147–161
    bartering and, 151
    cleaning with vinegar, 160–161
    foraging, 152
    herbs and natural remedies,
      154–160
    repurposing and reclaimed
      materials, 148–150
    thrifting, 148
Rhode Island Reds, 37, 54
Rhodes, Justin, 40
Rhubarb, 109
Roaster oven, 172
Roosters, 34, 38, 42
Rosemary, 110

S

Saanen goats, 95, 96
Scratch (chicken), 42

Seeds
    direct-sowing *versus* starting
        indoors, 124–125
    saving, 126
Skills, learning new, 28
Soil, 20, 118–122
Soil tests, 20, 25, 118–119
Stanchion, 85
Storage, of meat, 75
Straw mulch, 122
Stress, chickens and, 46
Suburban areas, homesteading in, 18
Succession planting, 114–115
Sunchokes (Jerusalem artichokes),
    110
Sunflowers, 126
Supers (honey production), 130
Sustainability, 52, 72, 184. *See also*
    Resourceful living; Waste
Swarm (bee), 137, 139
Syrups, 156–157

T

Tamworths (pigs), 65
Tea, making from herbs, 155, 162
Thrifting/thrift shopping, 148
Toggenburg goats, 95, 96
Tomatoes, 124, 126, 168, 170, 174
Tools
    for bee keeping, 135–136
    for butchering and processing
        animals, 73
    for homesteader kitchen,
        168–172
Top bar hives, 132
Turkeys, 49
    breeding, 57
    breeds, 54, 55
    butchering, 72
    housing, 56

U

USDA growing zone, 105
USDA inspection, 73

V

Veal, 67
Veterinary care, 72
Vinegar, cleaning with, 160–161

W

Walking onions, 110
Waste
    chicken, 43
    eating leftovers and, 182
    industrial *vs.* pasture-based
        meat and, 52
    milk and, 100
    pig, 62
    rabbit, 60
Water
    garden site and, 106
    land shopping and source of,
        20
    for milking animals, 88
    siting a bee yard and, 140
Welsummers (chickens), 45
Wet feet, egg production and, 45
Whiting True Blue chickens, 45
Winter
    bee keeping and, 141
    egg laying in the, 32, 38, 45
    growing food in the, 115
Winter squash, 115, 124, 126, 173, 178
Wood chips, 122

Y

Yard birds. *See also* Chicken(s)
    ducks, 47–48
    geese, 48–49
    guineas, 46
    quail, 46–47
    turkeys, 49
Yorkshires (pigs), 65

Z

Zones, permaculture, 25–26, 106,
    140